“十三五”普通高等教育规划教材

信息安全英语教程

张强华　马　飞　司小侠　编著

机械工业出版社

本书选材广泛，覆盖了网络安全、无线网络安全、移动网络安全、数据库安全、数据备份、物联网安全、大数据与安全以及云计算安全等内容。

每个单元包含以下部分：课文——选材广泛、风格多样、切合实际的两篇专业文章；单词——给出课文中出现的新词，读者由此可以积累专业基本词汇；词组——给出课文中的常用词组；缩略语——给出课文中出现的、业内人士必须掌握的缩略语；注释讲解——讲解课文中出现的关键知识点，不仅释疑解惑，而且扩大了读者的阅读面，为读者进一步的学习提供思路，构建了网状的知识体系；习题——既有针对课文的练习，也有一些开放性的练习，力求丰富读者知识储备；阅读材料——进一步扩大读者的视野；参考译文（Text A）——让读者对照理解和提高翻译能力。本书课文配有音频材料，扫描二维码即可收听；扫描二维码还可以免费获得本书的词汇总表，既可用于复习和背诵，也可作为小词典随时查阅。

本书教学资源丰富，有配套的PPT、参考答案等资料，便于教师组织教学。

本书既可作为高等院校本科和专科信息安全相关专业的专业英语教材，也可供从业人员自学，还可作为培训班的培训用书。

图书在版编目（CIP）数据

信息安全英语教程 / 张强华，马飞，司小侠编著. —北京：机械工业出版社，2018.6

“十三五”普通高等教育规划教材

ISBN 978-7-111-60591-1

Ⅰ.①信… Ⅱ.①张… ②马… ③司… Ⅲ.①信息安全－英语－高等学校－教材 Ⅳ.①TP309

中国版本图书馆CIP数据核字（2018）第171529号

机械工业出版社（北京市百万庄大街22号 邮政编码100037）
责任编辑：郝建伟 杨 洋
责任印制：张 博
唐山三艺印务有限公司印刷

2018年9月第1版·第1次印刷
184mm×260mm·15印张·368千字
0001－3000册
标准书号：ISBN 978-7-111-60591-1
定价：49.00元

凡购本书，如有缺页、倒页、脱页，由本社发行部调换

电话服务
服务咨询热线：（010）88379833
读者购书热线：（010）88379649

网络服务
机 工 官 网：www.cmpbook.com
机 工 官 博：weibo.com/cmp1952
教育服务网：www.cmpedu.com
金 书 网：www.golden-book.com

前　言

信息领域的安全问题层出不穷，在未来一段可以预见的时间区间内，信息安全专业人才需求巨大。因此，许多高校开设了“信息安全”专业，培养社会急需的专业人才。而信息安全技术发展极为迅猛，要求从业人员必须掌握许多新技术、新方法，由此提升了对其专业英语水平的要求。在职场中，具备相关职业技能并精通专业外语的人员往往能赢得竞争优势，成为不可或缺的甚至是引领性的人才。本书旨在帮助读者提高信息安全专业英语水平。

本书具有如下特色：

（1）选材广泛。本书覆盖了网络安全、无线网络安全、移动网络安全、信息安全与信息安全系统、风险管理、黑客、有害软件、拒绝服务攻击、防火墙、反病毒软件、入侵检测系统、端口扫描程序、加密和解密、密码保护、公钥基础设施、数字认证和数字签名、网站安全、数据库安全、数据备份、物联网安全、大数据与安全以及云计算安全等内容。

（2）体例创新，适合教学。本书的内容设计与课堂教学的各个环节紧密切合。每个单元都包含以下部分：课文——提供选材广泛、风格多样、切合实际的两篇专业文章；单词——给出课文中出现的新词，读者由此可以积累专业基本词汇；词组——给出课文中的常用词组；缩略语——给出课文中出现的、业内人士必须掌握的缩略语；注释讲解——讲解课文中出现的关键知识点，不仅释疑解惑，而且扩大了读者的阅读面，为读者进一步的学习提供思路，构建了网状的知识体系；习题——既有针对课文的练习，也有一些开放性的练习，力求丰富读者知识储备；阅读材料——进一步扩大读者的视野；参考译文（Text A）——让读者对照理解和提高翻译能力。本书课文配有音频材料，扫描相应的二维码下载后即可收听；扫描封底二维码还可以下载本书的词汇总表，既可用于复习和背诵，也可作为小词典随时查阅。

（3）教学资源丰富，教学支持完善。本书有配套的 PPT、参考答案等资料。另外，书中的习题量适当，题型丰富，难易搭配，便于教师组织教学。

本书的编者具有 28 年的行业英语翻译、教学与图书编写经验，编者力求使本书更符合教学要求，让教师和学生使用本书时更得心应手。在本书使用过程中，有任何问题，读者都可以通过电子邮件与我们交流，我们一定会及时答复。我们的 E-mail 地址如下：zqh3882355@sina.com；zqh3882355@163.com。教师也可到机械工业出版社网站免费下载课件。

本书既可作为高等院校本科和专科信息安全相关专业的专业英语教材，也可供从业人员自学，还可作为培训班的培训用书。

编　者

目　　录

Unit 1

Text A

Network Security

Network security consists of the policies and practices adopted to prevent and monitor unauthorized access, misuse, modification, or denial of a computer network and network-accessible resources. Network security involves the authorization of access to data in a network, which is controlled by the network administrator. Users choose or are assigned an ID and a password or other authentication[1] information that allows them access to information and programs within their authority. Network security covers a variety of computer networks, both public and private, that are used in everyday jobs; conducting transactions and communications among businesses, government agencies and individuals. Networks can be private, such as within a company, and others which might be open to public access. Network security is involved in organizations, enterprises, and other types of institutions. It does as its title explains: It secures the network, as well as protecting and overseeing operations being done. The most common and simple way of protecting a network resource is by assigning it a unique name and a corresponding password.

Network security starts with authentication, commonly with a username and a password. Since this requires just one detailed authentication, this is sometimes termed one-factor authentication. With two-factor authentication[2], something the user "has" is also used (e.g., a security token[3] , an ATM card, or a mobile phone); and with three-factor authentication, something the user "is" also used (e.g., a fingerprint or retinal scan[4]).

Once authenticated, a firewall enforces access policies such as what services are allowed to be accessed by the network users. Though effective to prevent unauthorized access, this component may fail to check potentially harmful content such as computer worms or Trojans being transmitted over the network. Antivirus software or an Intrusion Prevention System (IPS) help detect and inhibit the action of such malware. An anomaly-based intrusion detection system may also monitor the network like Wireshark[5] traffic and may be logged for audit purposes and for later high-level analysis. Newer systems combining unsupervised machine learning with full network traffic analysis can detect active network attackers from malicious insiders or targeted external attackers that have compromised a user machine or account.

Communication between two hosts using a network may be encrypted to maintain privacy.

Honeypots[6], essentially decoy network-accessible resources, may be deployed in a network as surveillance and early warning tools, as the honeypots are not normally accessed for legitimate purposes. Techniques used by the attackers that attempt to compromise these decoy resources are studied during and after an attack to keep an eye on new exploitation techniques. Such analysis may be used to further tighten security of the actual network being protected by the honeypot. A honeypot can also direct an attacker's attention away from legitimate servers. A honeypot encourages attackers to spend their time and energy on the decoy server while distracting their attention from the data on the real server. Similar to a honeypot, a honeynet is a network set up with intentional vulnerabilities. Its purpose is also to invite attacks so that the attacker's methods can be studied and that information can be used to increase network security. A honeynet typically contains one or more honeypots.

With all of the vital personal and business data being shared on computer networks every day, security has become one of the most essential aspects of networking. No one recipe to fully safeguard networks against intruders exists. Network security technology improves and evolves over time as the methods for both attack and defense grow more sophisticated.

1. Physical Network Security

The most basic but often overlooked element of network security involves keeping hardware protected from theft or physical intrusion.

Corporations spend large sums of money to lock their network servers, network switches and other core network components in well-guarded facilities. While these measures aren't practical for homeowners, households should still keep their broadband routers in private locations, away from nosy neighbors and house guests.

The widespread use of mobile devices makes physical security much more important. Small gadgets are especially easy to leave behind at travel stops or to have fall out of pockets.

Finally, stay in visual contact with a phone when loaning it to someone else: A malicious person can steal personal data, install monitoring software, or otherwise "hack" phones in just a few minutes when left unattended.

2. Password Protection

Passwords are an extremely effective system for improving network security if applied properly. Unfortunately, some don't take password management seriously and insist on using bad, weak (meaning, easy to guess) passwords like "123456" on their systems and networks.

Following just a few common-sense best practices in password management will greatly improve the security protection on a computer network:

- Set strong passwords, or passcodes, on all devices that join the network.
- Change the default administrator password of network routers.
- Do not share passwords with others more often than necessary; set up guest network access for friends and visitors if possible; change passwords when they may have become too widely known.

3. Spyware

Even without physical access to the devices or knowing any network passwords, illicit programs called spyware can infect computers and networks, typically by visiting websites. Much spyware exists on the Internet. Some spyware monitors a person's computer usage and Web browsing habits and reports this information back to corporations, who use it to create more targeted advertising. Other spyware attempts to steal personal data. One of the most dangerous forms of spyware, keylogger software[7], captures and sends the history of all keyboard keystrokes a person makes, ideal for capturing passwords and credit card numbers. All spyware on a computer attempts to function without the knowledge of people using it, thereby posing a substantial security risk.

Because spyware is notoriously difficult to detect and remove, security experts recommend installing and running reputable anti-spyware software on computer networks.

4. Online Privacy

Personal stalkers, identity thieves and perhaps even government agencies monitor people's online habits and movements well beyond the scope of basic spyware. WiFi hotspot usage from commuter trains and automobiles reveal a person's location, for example. Even in the virtual world, much about a person's identity can be tracked online through the IP addresses of their networks and their social network activities.

Techniques to protect a person's privacy online include anonymous Web proxy servers, although maintaining full privacy online is not fully achievable through today's technologies.

New Words

network	*n.*	网络
policy	*n.*	政策，方针
practice	*n.*	实行，实践，实际，惯例，习惯
prevent	*v.*	防止，预防
unauthorized	*adj.*	未被授权的，未经认可的
access	*n.*	访问，入门
	v.	存取
denial	*n.*	否认，否定，拒绝
resource	*n.*	资源
authorization	*n.*	授权，认可
data	*n.*	数据
control	*n. & v.*	控制，支配，管理，操纵
administrator	*n.*	管理员
assign	*v.*	分配，指派
password	*n.*	密码，口令
authenticate	*v.*	鉴别
program	*n.*	程序，计划
authority	*n.*	权威，威信

cover	*v.*	包括，包含；适用
communication	*n.*	通信
individual	*n.*	个人，个体
	adj.	个别的，单独的，个人的
enterprise	*n.*	企业
institution	*n.*	公共机构，协会
explain	*v.*	解释，说明
oversee	*v.*	监视，检查
username	*n.*	用户名
token	*n.*	令牌
fingerprint	*n.*	指纹，手印
	v.	采指纹
firewall	*n.*	防火墙
component	*n.*	成分
	adj.	组成的，构成的
Trojan	*n.*	特洛伊木马
transmit	*v.*	传输，传播
detect	*v.*	检测，探测，察觉，发觉，侦查
inhibit	*v.*	抑制，约束
malware	*n.*	恶意软件，流氓软件
anomaly	*n.*	不正常的
traffic	*n.*	流量，通信量
unsupervised	*adj.*	无人监督的，无人管理的
active	*adj.*	积极的，主动的
attacker	*n.*	攻击者
malicious	*adj.*	怀恶意的，恶毒的
insider	*n.*	内部的人，知道内情的人，权威人士
compromise	*v.*	危及……的安全
account	*n.*	账号
encrypt	*v.*	加密，将……译成密码
maintain	*v.*	维持，维护
decoy	*v.*	诱骗
surveillance	*n.*	监视，监督
legitimate	*adj.*	合法的，合理的
	v.	合法
exploitation	*n.*	开发，利用，剥削
encourage	*v.*	鼓励
distract	*v.*	转移
honeynet	*n.*	蜜网

vulnerability	*n.*	弱点；攻击
recipe	*n.*	处方
safeguard	*v.*	维护，保护，捍卫
	n.	安全装置，安全措施
hardware	*n.*	硬件
intrusion	*n.*	闯入，侵扰
server	*n.*	服务器
switch	*n.*	交换机
broadband	*n.*	宽带
router	*n.*	路由器
nosy	*adj.*	好管闲事的，爱追问的
	n.	好管闲事的人
widespread	*adj.*	分布广泛的，普遍的
device	*n.*	装置，设备
properly	*adv.*	适当地，完全地
unfortunately	*adv.*	不幸地
passcode	*n.*	一次性密码
default	*n.*	默认(值)，缺省(值)
illicit	*adj.*	违法的
spyware	*n.*	间谍软件
habit	*n.*	习惯，习性
dangerous	*adj.*	危险的
capture	*v.*	捕获
substantial	*adj.*	实质的，真实的，充实的
notorious	*adj.*	声名狼藉的
remove	*v.*	删除，移去
reputable	*adj.*	著名的
stalker	*n.*	跟踪者
identity	*n.*	身份
activity	*n.*	行动，行为
anonymous	*adj.*	匿名的
achievable	*adj.*	做得成的，可完成的

Phrases

consist of	由……组成
computer network	计算机网络
a variety of	多种的
government agency	政府机构
start with…	以……开始

one-factor authentication	单因素认证，单身份验证
two-factor authentication	双因素认证；双重身份验证
three-factor authentication	三因素认证；三重身份验证
retinal scan	虹膜扫描
fail to	未能
computer worm	计算机蠕虫
antivirus software	抗病毒软件，防病毒软件
machine learning	机器学习
early warning	预警
keep an eye on	密切注视，照看
protect from	保护
network component	网络元件
away from	远离
insist on	坚持
keylogger software	键盘记录软件
WiFi hotspot	WiFi 热区
social network	社交网络
proxy server	代理服务器

Abbreviations

ID (identification, identity)	身份，标识符
IPS (Intrusion Prevention System)	预防入侵系统
WiFi (Wireless Fidelity)	基于 IEEE 802.11b 标准的无线局域网
IP (Internet Protocol)	网际协议

Notes

[1] Authentication is the act of confirming the truth of an attribute of a single piece of data claimed true by an entity. In contrast with identification, which refers to the act of stating or otherwise indicating a claim purportedly attesting to a person or thing's identity, authentication is the process of actually confirming that identity.

[2] Two-factor authentication (also known as 2FA) is a method of confirming a user's claimed identity by utilizing a combination of two different components. Two-factor authentication is a type of multi-factor authentication.

[3] Security tokens are used to prove one's identity electronically (as in the case of a customer trying to access their bank account). The token is used in addition to or in place of a password to prove that the customer is who they claim to be. The token acts like an electronic key to access something.

[4] A retinal scan is a biometric technique that uses the unique patterns on a person's retina blood vessels.

[5] Wireshark is a free and open source packet analyzer. It is used for network troubleshooting, analysis, software and communications protocol development, and education.

[6] In computer terminology, a honeypot is a computer security mechanism set to detect, deflect, or, in some manner, counteract attempts at unauthorized use of information systems. Generally, a honeypot consists of data (for example, in a network site) that appears to be a legitimate part of the site but is actually isolated and monitored, and that seems to contain information or a resource of value to attackers, which are then blocked. This is similar to the police baiting a criminal and then conducting undercover surveillance, and finally punishing the criminal.

[7] A keylogger is a software that records the real time activity of a computer user including the keyboard keys they press.

Exercises

[Ex. 1] Answer the following questions according to the text.

1. What does network security consist of?
2. What does network security involve?
3. What is the most common and simple way of protecting a network resource?
4. What does a honeypot encourage attackers to do?
5. What is the purpose of a honeynet?
6. What does the most basic but often overlooked element of network security involve?
7. What are the few common-sense best practices in password management?
8. What is one of the most dangerous forms of spyware mentioned in the text? What does it do?
9. Why do security experts recommend installing and running reputable anti-spyware software on computer networks?
10. How can much about a person's identity be tracked online even in the virtual world?

[Ex. 2] Translate the following terms or phrases from English into Chinese and vice versa.

1. malicious	1.
2. program	2.
3. switch	3.
4. machine learning	4.
5. proxy server	5.
6. 攻击者	6.
7. 鉴别	7.
8. 匿名的	8.
9. 默认(值)，缺省(值)	9.
10. 授权，认可	10.

[Ex. 3] Translate the following passage into Chinese.

People are more active online than ever, which can mean increased risk of having your data exploited. But there are simple steps everyone can take to minimize danger.

As a trained risk manager I see lots of data regarding how careless individuals are when they venture into cyberspace. People love to login to their social networks and post pictures, let everyone know where they are going to be and when, then tell the world of their travels and/or shopping sprees. Many people out there don't even attempt to hide their personal information from the general public—their "friends"—let alone the various criminals and hackers who are swarming the Internet every second of every day. And that isn't the half of it. We shop online, too, with unfettered abandon, making ourselves open books for any and all to see. Back in the old days, the bad guys had to use a gun to get our wallets and steal our credit cards. Now, we basically give it to them in the form of self-created peepholes in our Internet privacy.

[Ex. 4] Fill in the blanks with the words given below.

card	address	information	spyware	Internet
smartest	malicious	sold	password	connection

Six Free Ways to Take Control of Your Internet Privacy

1. Clean Up Your Cyber Footprint

Get a copy of your credit report and use this as your foundation to correct any misinformation regarding past delinquencies and already closed accounts. Close any and all social media services that you don't use, and think twice before sharing your personal ___1___ with anyone on any platform. At a minimum, make sure your ___2___ is secure (encrypted) by looking for HTTPS:// as opposed to HTTP:// at the beginning of your intended Web address.

2. Isolate E-mail and Electronic Payment Methods

If you are going to buy online, and most of us do, the ___3___ thing to do is consolidate your cyber tools. Use a designated e-mail account for ecommerce, and most importantly, if you insist on using plastic, use a dedicated credit ___4___. However, it is always recommended to use an electronic option such as PayPal, Apple Pay or Amazon Payments.

3. Use a Password Generator

Stop using the same weak password—12345, your middle name or a child's birthday. Companies including Dashlane.com and Keepass.info offer free ___5___ generators and online password vaults that are very effective in protecting you and your personal information from the point of login.

4. Use Caution

Be proactive when browsing the net as some pages (porn, music, file-sharing sites, etc.) are more apt than others to have ___6___, malware and/or hackers lurking about, and there is no way to know when or where you are at risk. Consider taking advantage of free anti-malware and/or anti-virus programs from Microsoft, Malwarebytes, AVG, et al. Learn how to control your cookie intake by adjusting your security-related Internet settings.

5. Heed Warnings

Pop-ups occur for a reason. Don't give out personally identifiable information too easily. Just as you might think twice about giving some clerk at the mall your home ___7___ and phone number, keep in mind that simply because a site asks for or demands personal information from you does not mean you have to comply.

6. Don't Engage

We have all been told "If you don't have anything nice to say..." That old adage most definitely applies to the ___8___, with the emphasis shifting to written speech instead of spoken words. Point is, if you don't have anything nice to write, don't. Don't engage in blogs, chat sessions or other forums where misinformation, hate or other ___9___ endeavors may be the intent. If you think what you are doing is hurtful, it probably is. Never lash out or try to hurt other people's feelings with your comments online. Remember your written words never go away once posted. They're trackable and traceable, like a trail of breadcrumbs that can lead back to you months or years after you no longer care or feel differently about the subject or person.

Regarding Terms of Use and Privacy Agreements: All social networks and ecommerce sites have Terms of Use and Privacy Agreements. Don't be shy about not signing up for a site because you don't like how they may expose you. Many sites can and do commoditize your data. The only hope that your information won't be ___10___ is if you are paying for the site or service, and even then they may sell you out. Set your accounts to their highest privacy settings. There are a lot of bad people out there. The more streamline you route, the safer you and your personal information will be.

Text B

What You Should Know about Your Privacy Online

Every day it seems like we lose a little more anonymity. Everywhere we go, people want a taste of our private information. Whether you're checking in at the airport, signing up for a gym membership or paying bills at the bank, you can't do any of these things without giving up a few scraps of personal data. The same goes for making use of all the great sites and services offered by the Internet. Whether you're using a free e-mail account, like Gmail or Hotmail, trawling eBay for deals or signing up to find a bit of companionship on a site like Plentyoffish.com, you are expected to provide any number of personal details.

But how is the information you provide online used? The answer might surprise you.

1. What Information Is Shared about You Online

Before we can address the ways your personal information is used—and sometimes abused—online, let's talk about the two different kinds of information that online service providers, retailers and websites collect from users when they access the Internet.

(1) Personally Identifiable Information (PII[1])

PII can include data, such as your name, address, banking credentials and even your Internet access location—be it your personal computer or on the go with a tablet or smartphone.

(2) Non-Personally-Identifiable Information (NPII)

The second, and somewhat less menacing, type of data collected as a result of your time online is known as NPII, which is when a website counts how many visitors it attracts on a daily basis, or what site content is clicked the most by users. The difference here is that the website is not interested in individual users, but how they behave collectively.

PII and NPII are collected online through a number of methods. In some cases, users knowingly hand over their personal information. Providing an online retailer with your delivery address, uploading a location-tagged photo to a service like Flickr[2], or checking in with a service like Foursquare[3] are great examples of this. When it comes to NPII, one of the most popular methods for collecting data is through the use of HTTP cookies—data stored on a site visitor's computer, smartphone or tablet that provide the website with the user's preferences, such as the contents of an online shopping cart, or the settings associated with a user's previous visits.

2. Who's Using Your Personal Information Online

In the same way that a firearm can be wielded as either a tool or a weapon, your personal information can be used to benefit you and/or put you at risk. Some services and websites use PII and NPII to enrich your time online. For example, when you go shopping at Amazon.com, Amazon retains information about your searched and purchased products and uses that data to suggest other products of interest. Provided you're able to exercise a bit of willpower and not click your way into the poorhouse every time Amazon serves up something new and tantalizing, such suggestions are a great way to discover new products that you may well end up loving.

That said, it's also possible for websites, hackers and other online ne'er-do-wells to use your personal data to make your life—online and offline—a living hell. For example, the personal contact information required to create a profile on an online forum can be sold to Internet marketers who may use that information to spam your e-mail account with unwanted offers detailing dubious-looking products or tips on how to a buy into a Nigerian prince's discarded fortune. In extreme cases, users have entrusted personal information to questionable web services or insecure websites—only to discover that criminals used their information to hijack their identities, online accounts or even procure new credit cards—all things that are guaranteed to leave you feeling angry and violated. Undoing them is also a huge inconvenience and can take years.

3. How to Protect Your Online Privacy

So does protecting your personal information mean you have to limit your Internet exposure to a few terse e-mails? Or avoid shopping, social media, forums and any other services that ask for information about you? Not if you're careful. The following are a few tips even the least web-savvy Internet denizens can follow to ensure safe surfing.

- Only frequent sites that offer a clearly defined privacy policy that outlines how the personal information you submit will be used.
- Use a "throw away" e-mail address or alias when posting to forums or signing up for web services with questionable privacy policies. This will help you protect yourself from spam and other unwanted solicitations.
- Never submit highly sensitive personal information, such as your home address or credit card number, to a site unless you know it is secure/encrypted.

- Only use web browsers that support private browsing. These include Safari, Firefox or Chrome. Private browsing prevents your browser from storing or transmitting your browsing habits to offsite services.

4. What They Know

The notion of online privacy is a tricky subject. While news of new privacy rules on Facebook or Google often gets users up in arms, the fact is that while there is a lot of information shared about you over the web, most of it is not personal. That isn't to say that invasions of personal privacy do not happen, but most websites actually collect data to improve user experience and boost business—not to scam you. The good news is that when it comes to real fraud, you can often protect yourself with a little know-how and common sense.

New Words

rule	*n.*	规则，惯例，准则，标准
	v.	规定；统治，支配
anonymity	*n.*	匿名，作者不明(或不详)
gym	*n.*	体育馆；体操
scrap	*n.*	小片
	adj.	零碎的，废弃的
e-mail	*n.*	电子邮件
deal	*n.*	交易，买卖
	v.	处理，应付，做生意；分配，给予
website	*n.*	网站
collect	*v.*	收集，聚集，集中，搜集
Internet	*n.*	互联网
menace	*n.*	威胁，危险物
	v.	恐吓，危及，威胁
attract	*v.*	吸引；有吸引力，引起注意
click	*v.*	（用鼠标）点击
upload	*v.*	上传
firearm	*n.*	火器，枪炮
enrich	*v.*	节省
suggest	*v.*	建议，提出
offline	*adj.*	未联机的，脱机的，离线的
	adv.	未联机地，脱机地，离线地
forum	*n.*	论坛
dubious	*adj.*	可疑的，不确定的
discard	*v.*	丢弃，抛弃
entrust	*v.*	委托
questionable	*adj.*	可疑的

insecure	*adj.* 不可靠的，不安全的
criminal	*n.* 罪犯，犯罪者
	adj. 犯罪的，犯法的，罪恶的
procure	*v.* 获得，取得
angry	*adj.* 生气的，愤怒的
inconvenience	*n.* 麻烦，不方便之处
terse	*adj.* 简洁的，扼要的
surfing	*n.* 网络冲浪
submit	*v.* 提交，递交
alias	*n.* 别名，化名
solicitation	*n.* 恳求，恳请，诱惑，游说，征稿
offsite	*n.* 离站，异地
tricky	*adj.* 狡猾的，机警的
boost	*v.* 推进
know-how	*n.* 技术秘诀，诀窍

Phrases

a taste of	一点，少量，小量
check in	登机验票，登记住宿，报到
sign up for	注册，选课；报名参加
give up	把……让给；放弃
make use of	使用，利用
product of interest	感兴趣的产品
a bit of	一点
online service provider	在线服务提供商
banking credential	银行凭证
on the go	方便携带的；忙个不停的，活跃的
as a result of	作为结果
be interested in…	对……有兴趣
a number of	许多的
hand over	交出，交付，交给，让与
delivery address	送货地址，交货地址；收信人地址
online shopping cart	在线购物车
be wielded as	可充当
ne'er-do-wells	从没干好事，从不干好事(=never-do-wells)
social media	社交媒体
throw away	扔掉，丢弃
get sb. up in arms	使某人满腔怒火，竭力反对
common sense	常识

Abbreviations

PII (Personally Identifiable Information) 个人验证信息

NPII (Non-Personally-Identifiable Information) 非个人验证信息

Notes

[1] Personally Identifiable Information (PII), or Sensitive Personal Information (SPI), as used in information security and privacy laws, is information that can be used on its own or with other information to identify, contact, or locate a single person, or to identify an individual in context.

[2] Flickr is an image hosting and video hosting website and web services suite that was created by Ludicorp in 2004 and acquired by Yahoo on March 20, 2005. In addition to being a popular website for users to share and embed personal photographs, and effectively an online community, the service is widely used by photo researchers and by bloggers to host images that they embed in blogs and social media.

[3] Foursquare is a local search-and-discovery service mobile app which provides search results for its users. The app provides personalized recommendations of places to go to near a user's current location based on users' "previous browsing history, purchases, or check-in history". Foursquare lets the user search for restaurants, nightlife spots, shops and other places of interest in their surrounding area. It is also possible to search other areas by entering the name of a remote location. The app displays personalised recommendations based on the time of day, displaying breakfast places in the morning, dinner places in the evening etc. Recommendations are personalised based on factors that include a users check-in history, their "Tastes" and their venue ratings and according to their friends reviews.

Exercises

[Ex. 5] Answer the following questions according to the text.

1. What is the main idea of the first paragraph?
2. What does PII stand for?
3. What can PII include?
4. What does NPII stand for?
5. What is one of the most popular methods for collecting data when it comes to NPII?
6. What is the example given in the passage to show that your personal information can be used to benefit you?
7. What is the example given in the passage to show that your personal information can be used to put you at risk?
8. What happens when users have entrusted personal information to questionable web services or insecure websites in extreme cases?

9. Why does the author suggest using a "throw away" e-mail address or alias when posting to forums or signing up for web services with questionable privacy policies?

10. Why does the author suggest only using web browsers that support private browsing?

Reading Material

Top Must-Know Network Security Tricks

For new network administrators, there are a few security tricks that can keep your infrastructure safe from attacks and malicious behavior. Below are tips and recommendations[1] from industry experts who know the importance of securing your network from internal and external intrusions.

1. Create Unique[2] Infrastructure Passwords

"One of the most common mistakes made in network security is to implement a shared password across all routers, switches, and infrastructure devices. The compromise of one device or leakage[3] of the password can lead to a compromise of every device—not only by an insider, but also by malware that targets these devices. As a best practice security recommendation, each device should have a unique password in order to limit the liability of the device being leveraged[4] against its peers."

—Morey Haber, VP of Security, BeyondTrust

2. Use Scheduled and Event Based Password Changes

"While the management of network passwords has become commonplace with a variety password management tools, the changing of passwords on a large scale[5] generally requires a dedicated 'password safe' tool. Manually changing passwords for hundreds or thousands of devices is labor intensive and generally avoided by organizations because of time constraints, unless a dedicated[6] tool is procured. As per regulatory compliance initiatives and security best practices, passwords should always be rotated on a periodic[7] basis, making this exercise rather problematic[8]. In addition, based on events, ad-hoc changes may need to occur to passwords based on employment changes and events like contractor access.

Therefore, it is recommended to use a schedule to change all of your network passwords on a regular basis, or have a process to change them ad-hoc if needed, and if the time required to make

---

1 recommendation　*n.*　推荐，劝告，建议

2 unique　*adj.*　唯一的，独特的

3 leakage　*n.*　漏，泄漏

4 leverage　*v.*　利用，举债经营

5 on a large scale　大规模地

6 dedicated　*adj.*　专门的

7 periodic　定期的，周期的

8 problematic　*adj.*　问题的，有疑问的

these changes is excessive[1], consider a solution to automate[2] the process."

—Morey Haber

3. Harden the Devices from Default Settings

"Almost every device has default settings[3] and passwords when it is first installed. It is up to the end user to change account names, passwords, secure ports[4], and harden the device from malicious activity.

As a best practice[5] for network security, it is recommended to change all these settings, and to also use a tool to assess whether the devices are properly hardened, do not contain default passwords for things like SNMP[6], and are running the latest firmware[7] to vet out any vulnerabilities and missing security patches[8]. Unfortunately, many organizations install devices and do not actively place them into a device management life cycle for patching, configuration[9], and maintenance to keep them secure like servers or workstations[10]."

—Morey Haber

4. Set Up a Dedicated Network for Business-Critical Servers

"Segregating[11] business critical servers and systems to their own network or subnetwork[12] can often be very beneficial[13]. By placing these systems in their own network you maximize[14] their connection quality and decrease their network latency[15] by limiting the amount of non-critical traffic near them. Additionally, you gain the security of logically separating the packets around your most important systems allowing you to better monitor and shape the network traffic."

—Neil Marr, Director of IT, LearnForce Partners LLC, ExamForce

5. Always Log and Study Suspicious[16] Activity

"Deny all outbound[17] network traffic not specifically allowed and log it. Then look at it. The log data will provide an enormous[18] amount of information about what is happening, both good and bad, on your network."

---

1 excessive *adj.* 过多的，过分的，额外的
2 automate *v.* 使自动化，自动操作
3 default setting 默认设置
4 port *n.* 端口
5 best practice 最优方法
6 SNMP (Simple Network Management Protocol) 简单网络管理协议
7 firmware *n.* 固件，韧件(软件硬件相结合)
8 patch *n.* 补丁
9 configuration *n.* 构造，配置
10 workstation *n.* 工作站
11 segregate *v.* 隔离
12 subnetwork *n.* 子网络
13 beneficial *adj.* 有益的，受益的
14 maximize *v.* 取……最大值，最佳化
15 latency *n.* 潜伏，潜在；反应时间
16 suspicious *adj.* 可疑的，怀疑的
17 outbound *adj.* 出站的，出去的
18 enormous *adj.* 巨大的，庞大的

—Brian O'Hara, Senior Security Consultant, Rook Security

6. Be Picky about Your Traffic

"Only allow specific, required traffic both inbound[1] and outbound on your network. If an application and/or associated port is not needed, block[2] it. This takes some time but is well worth the effort."

—Brian O'Hara

7. Block IP Ranges from Suspicious Foreign Bodies

Unless you do business in Russia, DPRK(the Democratic People's Republic fo Korea), etc. block their IP ranges. It will not stop those that are really interested in getting to your facilities but it will stop a lot of "browsing around" to see what you have.

—Brian O'Hara

8. Never Use Default Security Settings on Internet Devices

"Make sure you have hardened any Internet facing device and change the Admin username and password to something fairly hard to crack!"

—Brian O'Hara

9. Break the Addiction[3]

"Assume your user base is compromised and break your addiction to big data analytics in your security program. Today, most security strategies focus on protecting everything—including the user host. This creates volumes of data that burdens[4] artificial intelligence[5].

Assume your user base is compromised and focus your security efforts in anomaly detection between users, applications and the database. Some may dispute[6] that this is possible, but the banking industry has achieved this with their customer base. Enterprises can mirror this approach with success."

—Jeff Schilling, Chief Security Officer, Armor

10. Flip the Business Script

"Your cyber security problem may actually be a flawed[7] business process that was never engineered for automation or security. Case in point: How we process credit card payments. For the most part, this process remains based on the same principles used when we required physical paper-based card swiping that would be sent through card-processing chain.

"Examine your business processes and identify game-changing controls that will have a huge return on investment[8]. When gas stations began requiring ZIP codes at the pump, point-of-sale fraud dropped drastically. In fact, I know of one gas station chain where this change reduced credit card fraud by 94 percent."

---

1 inbound *adj.* 入站的，进入的
2 block *v.* 阻塞
3 addiction *n.* 沉溺，上瘾
4 burden *v.* 负担
5 artificial intelligence 人工智能
6 dispute *v.* 争论，辩论
7 flawed *adj.* 有缺陷的，有瑕疵的
8 return on investment 投资回报

—Jeff Schilling

11. Elevate & Control

"Implement permission access management (PAM)[1] controls that only authorize elevated[2] privileged access for a short time. Most APTs[3] are successful because they obtain elevated privileges[4], which they keep indefinitely. This allows threat actors to holster their arsenal of malicious tools and "live off the land" as an authorized user with elevated privileges.

There are many PAM solutions that allow organizations to manage elevated, time-based privileges. Some may even be mapped back to a trouble ticket[5]. This breaks the cyber kill chain for threat actors and stops them from being persistent[6]."

—Jeff Schilling

参考译文

网络安全

网络安全包括为防止和监控未经授权的访问、滥用、修改或拒绝计算机网络和可访问网络资源而采取的政策和做法。网络安全涉及访问网络中数据的授权，这由网络管理员控制。用户选择或被分配一个 ID 和密码或其他认证信息，让他们能在其权限内访问信息和程序。网络安全包括在日常工作中使用的各种公共和私人计算机网络，也包括进行企业、政府机构和个人之间的业务和通信。网络可以是私有的（如公司内的网络）也可能是对公众开放的网络。网络安全涉及组织、企业和其他类型的机构。如"网络安全"这一标题所示，它保护网络，并且保护和监控正在进行的操作。保护网络资源的最常见和简单的方法是通过为其分配唯一的名称和相应的密码而实现的。

网络安全始于认证，通常使用用户名和密码。由于这仅需要一个详细的用户名认证，因此也称为单因素认证。网络安全也能使用双因素认证，也就是使用用户"拥有"的东西（如安全令牌、ATM 卡或移动电话）。网络安全还可以使用三因素认证，用户还需要使用某些东西（如指纹或虹膜扫描）证明他们"是"谁。

一旦认证，防火墙就执行访问政策，例如允许网络用户访问什么服务。虽然这样做可以有效防止未经授权的访问，但此组件可能无法检查或许有害的内容（例如通过网络传输的计算机蠕虫或木马）。防病毒软件或入侵防御系统（IPS）有助于检测和禁止此类恶意软件的操

---

1 permission access management (PAM)　许可访问管理

2 elevated　*adj.*　升高的，高层的

3 APT (Advanced Persistent Threats)　高级持续性威胁

An advanced persistent threat is a set of stealthy and continuous computer hacking processes, often orchestrated by a person or persons targeting a specific entity. An APT usually targets either private organizations, states or both for business or political motives. APT processes require a high degree of covertness over a long period of time. The "advanced" process signifies sophisticated techniques using malware to exploit vulnerabilities in systems. The "persistent" process suggests that an external command and control system is continuously monitoring and extracting data from a specific target. The "threat" process indicates human involvement in orchestrating the attack.

4 privilege　*n.*　特权

5 trouble ticket　故障单，故障报告表

6 persistent　*adj.*　持久稳固的

作。基于异常入侵检测系统还可以监视网络（如 Wireshark 流量），并且可以记录下来用于以后的审计和高级分析。把无监督机器学习与全网络流量分析相结合的新系统可以检测主动的网络攻击者，这些攻击者来自内部或外部，目的在于恶意破坏用户的计算机或账户。

可以通过加密两个网络主机之间的通信来保持隐私。

蜜罐，其本质是作为诱饵的网络可访问资源，它可以部署在网络中作为监视和预警工具，因为蜜罐通常不用于合法目的的访问。通过在攻击期间和攻击后研究攻击者所使用的尝试破坏这些诱饵资源的技术，可以关注新的攻击技术。这样的分析可以用于进一步巩固由蜜罐保护的实际网络的安全性。蜜罐还可以将攻击者的注意力从合法的服务器上引开。蜜罐鼓励攻击者将其时间和精力花费在诱饵服务器上，同时分散他们对真实服务器上的数据的注意力。与蜜罐一样，蜜网也是故意设置了漏洞的网络，其目的也是吸引攻击，以便研究攻击者的方法，并且可以使用这些信息来提高网络安全性。蜜网通常包含一个或多个蜜罐。

由于所有重要的个人和商业数据每天都在计算机网络上共享，安全对网络而言至为重要。没有哪个方法可以完全保护网络，使之不受入侵。随着攻击和防御的方法越来越复杂，网络安全技术随着时间的推移也在不断改进和发展。

1. 物理网络安全

网络安全的最基本但经常被忽视的因素是保护硬件免受盗窃或物理入侵。

各类公司花费大量的资金将其网络服务器、网络交换机和其他核心网络部件保护在安全的设施中。虽然这些措施对于房主不切实际，但家庭仍应将其宽带路由器保留在私人地点，远离爱管闲事的邻居和客人。

移动设备的广泛使用使得物理安全变得更加重要。小物件特别容易忘在旅行站或从口袋里掉出来。

最后，当把手机借给别人时，应让其在视线之内：恶意的人可以偷窃个人数据、安装监控软件或者在“无人注意”时快速“黑掉”手机。

2. 密码保护

如果应用得当，密码就能够非常有效地提高网络的安全性。不幸的是，一些人不认真管理密码，并坚持在他们的系统和网络上使用不良的、弱的（就是容易猜到的）密码，如“123456”。

以下只是几个常用的密码管理方法，但可以大大提高计算机网络的安全保护：

- 在加入网络的所有设备上设置强密码或一次性密码。
- 更改网络路由器的默认管理员密码。
- 不要经常与他人共享密码；如果可能，就为朋友和来访者设置访客网络访问密码。如果知道这些密码的人较多，就更改密码。

3. 间谍软件

即使没有物理访问设备或不知道任何网络密码，称为间谍软件的非法程序也可以感染计算机和网络，通常访问网站时就会感染。互联网上存在大量间谍软件。某些间谍软件监视个人的计算机使用和网络浏览习惯，并将此信息报告给广告公司，以便这些公司使用它来创建更有针对性的广告。其他间谍软件尝试窃取个人数据。其中最危险的间谍软件形式之一是键盘监控软件，它捕获并发送一个人所有的键盘按键的历史记录，也是捕获密码和信用卡号码的理想工具。计算机上所有的间谍软件都试图在用户不知情的情况下工作，从而造成相当大的安全风险。

由于间谍软件非常难以检测和删除，网络安全专家建议在计算机网络上安装和运行著名的反间谍软件程序。

4. 在线隐私

个人追踪者、身份信息窃取者，甚至政府机构都在监控人们的网络习惯和行动，这远远超出了基本的间谍软件的范围。例如，使用来自通勤车和汽车的 WiFi 能够显示个人的位置。即使在虚拟世界中，通过网络的 IP 地址和社交网络活动，也可以在线跟踪很多的个人身份信息。

保护个人在线隐私的技术包括匿名 Web 代理服务器，但是当今的技术不能完全保护在线隐私。

Unit 2

Text A

Wireless Security

Wireless security is the prevention of unauthorized access or damage to computers using wireless networks. The most common types of wireless security are Wired Equivalent Privacy (WEP)[1] and WiFi Protected Access (WPA)[2]. WEP is a notoriously weak security standard. The password it uses can often be cracked in a few minutes with a basic laptop computer and widely available software tools. WEP is an old IEEE 802.11 standard from 1999, which was outdated in 2003 by WPA. WPA was a quick alternative to improve security over WEP. The current standard is WPA2. Some hardware cannot support WPA2 without firmware upgrade or replacement. WPA2 uses an encryption device that encrypts the network with a 256-bit key; the longer key length improves security over WEP.

Many laptop computers have wireless cards preinstalled, which allows people to enter a network while moving around. However, wireless networking is prone to some security issues. Hackers have found wireless networks relatively easy to break into, and even use wireless technology to hack into wired networks. As a result, it is very important that enterprises define effective wireless security policies that guard against unauthorized access to important resources. Wireless Intrusion Prevention Systems (WIPS)[3] or Wireless Intrusion Detection Systems (WIDS) are commonly used to enforce wireless security policies.

The risks to users of wireless technology have increased as the service has become more popular. There were relatively few dangers when wireless technology was first introduced. Hackers had not yet had time to latch on to the new technology, and wireless networks were not commonly found in the work place. However, there are many security risks associated with the current wireless protocols and encryption methods, and in the carelessness and ignorance that exists at the user and corporate IT level. Hacking methods have become much more sophisticated and innovative with wireless access. Hacking has also become much easier and more accessible with easy-to-use Windows- or Linux-based tools being made available on the web at no charge.

Some organizations that have no wireless access points installed do not feel that they need to address wireless security concerns. Issues can arise in a supposedly non-wireless organization when a wireless laptop is plugged into the corporate network. A hacker could sit out in the parking lot and gather information from it through laptops and/or other devices, or even break in through this wireless card-equipped laptop and gain access to the wired network.

1. Modes of Unauthorized Access

The modes of unauthorized access to links, to functions and to data is as variable as the respective entities make use of program code. There does not exist a full scope model of such threat. To some extent the prevention relies on known modes and methods of attack and relevant methods for suppression of the applied methods. However, each new mode of operation will create new options of threatening. Hence prevention requires a steady drive for improvement. The following described modes of attack are just a snapshot of typical methods and scenarios where to apply.

(1) Accidental Association

Violation of the security perimeter of a corporate network can come from a number of different methods and intents. One of these methods is referred to as "accidental association". When a user turns on a computer and it latches on to a wireless access point from a neighboring company's overlapping network, the user may not even know that this has occurred. However, it is a security breach in that proprietary company information is exposed and now there could exist a link from one company to the other. This is especially true if the laptop is also hooked to a wired network.

Accidental association is a case of wireless vulnerability called as "mis-association". Mis-association can be accidental, deliberate (for example, done to bypass corporate firewall) or it can result from deliberate attempts on wireless clients to lure them into connecting to attacker's APs.

(2) Malicious Association

"Malicious associations" are when wireless devices can be actively made by attackers to connect to a company network through their laptop instead of a company Access Point (AP). These types of laptops are known as "soft APs" and are created when a cyber criminal runs some software that makes his/her wireless network card look like a legitimate access point. Once the thief has gained access, he/she can steal passwords, launch attacks on the wired network, or plant trojans. Since wireless networks operate at the Layer 2 level, Layer 3 protections such as network authentication and Virtual Private Networks (VPNs)[4] offer no barrier. Wireless 802.1x authentications do help with some protection but are still vulnerable to hacking. The idea behind this type of attack may not be to break into a VPN or other security measures. Most likely the criminal is just trying to take over the client at the Layer 2 level.

(3) Ad Hoc Networks

Ad hoc networks can pose a security threat. Ad hoc networks are defined as peer to peer networks between wireless computers that do not have an access point in between them. While these types of networks usually have little protection, encryption methods can be used to provide security.

The security hole provided by ad hoc networking is not the ad hoc network itself. It is the bridge it provides into other networks, usually in the corporate environment and the unfortunate default settings in most versions of Microsoft Windows. This feature is turned on unless explicitly disabled. Thus the user may not even know they have an unsecured ad hoc network in operation on their computer. If they are also using a wired or wireless infrastructure network at the same time, they are providing a bridge to the secured organizational network through the unsecured ad hoc connection. Bridging is in two forms. One is direct bridge. It requires the user actually configure a bridge between the two connections and is thus unlikely to be initiated unless explicitly desired. The other is indirect bridge which is the shared resources on the user computer. The indirect bridge may expose private data that is shared from the user's computer to LAN connections, such as shared folders or private Network Attached Storage, making no distinction between authenticated or private connections and unauthenticated ad hoc networks.

(4) Non-traditional Networks

Non-traditional networks such as personal network Bluetooth[5] devices are not safe from hacking and should be regarded as a security risk. Even barcode readers, handheld PDAs, and wireless printers and copiers should be secured. These non-traditional networks can be easily overlooked by IT personnel who have narrowly focused on laptops and access points.

(5) Identity Theft (MAC Spoofing)

Identity theft (or MAC spoofing[6]) occurs when a hacker is able to listen in on network traffic and identify the MAC address of a computer with network privileges. Most wireless systems allow some kind of MAC filtering to allow only authorized computers with specific MAC IDs to gain access and utilize the network. However, programs exist that have network "sniffing" capabilities. Combine these programs with other software that allow a computer to pretend it has any MAC address that the hacker desires, and the hacker can easily get around that hurdle.

MAC filtering is effective only for small residential (SOHO) networks, since it provides protection only when the wireless device is "off the air". Any 802.11 device "on the air" freely transmits its unencrypted MAC address in its 802.11 headers, and it requires no special equipment or software to detect it. Anyone with an 802.11 receiver (laptop and wireless adapter) and a freeware wireless packet analyzer can obtain the MAC address of any transmitting 802.11 within range. In an organizational environment, where most wireless devices are "on the air" throughout the active working shift, MAC filtering provides only a false sense of security since it prevents only "casual" or unintended connections to the organizational infrastructure and does nothing to prevent a directed attack.

(6) Man-in-the-Middle Attacks

A man-in-the-middle attacker entices computers to log into a computer which is set up as a soft AP. Once this is done, the hacker connects to a real access point through another wireless card offering a steady flow of traffic through the transparent hacking computer to the real network. The hacker can then sniff the traffic. One type of man-in-the-middle attack relies on security faults in Challenge-Handshake Authentication Protocal(CHAP)[7] to execute a "de-authentication attack". This attack forces AP-connected computers to drop their connections and reconnect with the hacker's soft AP (disconnects the user from the modem so they have to connect again using their password which one can extract from the recording of the event). Man-in-the-middle attacks are enhanced by software such as LANjack and AirJack which automate multiple steps of the process, meaning what once required some skill can now be done by script kiddies[8]. Hotspots are particularly vulnerable to any attack since there is little to no security on these networks.

(7) Denial of Service (DoS)

A Denial-of-Service attack occurs when an attacker continually bombards a targeted AP or network with bogus requests, premature successful connection messages, failure messages, and/or other commands. These cause legitimate users to not be able to get on the network and may even cause the network to crash. These attacks rely on the abuse of protocols such as the Extensible Authentication Protocol (EAP)[9].

The DoS attack in itself does little to expose organizational data to a malicious attacker, since the interruption of the network prevents the flow of data and actually indirectly protects data by preventing it from being transmitted. The usual reason for performing a DoS attack is to observe the recovery of the wireless network, during which all of the initial handshake codes are re-transmitted by all devices, providing an opportunity for the malicious attacker to record these codes and use various cracking tools to analyze security weaknesses and exploit them to gain unauthorized access to the system. This works best on weakly encrypted systems such as WEP, where there are a number of tools available which can launch a dictionary style attack of "possibly accepted" security keys based on the "model" security key captured during the network recovery.

(8) Network Injection

In a network injection attack, a hacker can make use of access points that are exposed to non-filtered network traffic, specifically broadcasting network traffic such as "Spanning Tree" (802.1D), OSPF[10], RIP[11], and HSRP. The hacker injects bogus networking reconfiguration commands that affect routers, switches, and intelligent hubs. A whole network can be brought down in this manner and require rebooting or even reprogramming of all intelligent networking devices.

2. Wireless Intrusion Prevention Concepts

There are three principal ways to secure a wireless network:

1) For closed networks (like home users and organizations) the most common way is to configure access restrictions in the access points. Those restrictions may include encryption and checks on MAC address. Wireless Intrusion Prevention Systems can be used to provide wireless LAN security in this network model.

2) For commercial providers, hotspots and large organizations, the preferred solution is often to have an open and unencrypted, but completely isolated wireless network. The users will at first have no access to the Internet nor to any local network resources. Commercial providers usually forward all web traffic to a captive portal[12] which provides for payment and/or authorization. Another solution is to require the users to connect securely to a privileged network using VPN.

3) Wireless networks are less secure than wired ones; in many offices intruders can easily visit and hook up their own computer to the wireless network without problems, gaining access to the network, and it is also often possible for remote intruders to gain access to the network through backdoors like Back Orifice[13]. One general solution may be end-to-end encryption, with independent authentication on all resources that shouldn't be available to the public.

There is no ready designed system to prevent from fraudulent usage of wireless communication or to protect data and functions with wirelessly communicating computers and other entities. However, there is a system of qualifying the taken measures as a whole according to a common understanding what shall be seen as state of the art. The system of qualifying is an international consensus as specified in ISO/IEC 15408.

A Wireless Intrusion Prevention System (WIPS) is a concept for the most robust way to counteract wireless security risks. However such WIPS does not exist as a ready designed solution to implement as a software package. A WIPS is typically implemented as an overlay to an existing Wireless LAN infrastructure, although it may be deployed standalone to enforce no-wireless policies within an organization. WIPS is considered so important to wireless security that in July 2009, the Payment Card Industry Security Standards Council published wireless guidelines for PCI DSS recommending the use of WIPS to automate wireless scanning and protection for large organizations.

New Words

wireless	*adj.* 无线的
prevention	*n.* 预防，防止
damage	*n.* 损害，伤害
standard	*n.* 标准，规格
	adj. 标准的
available	*adj.* 可用到的，可利用的，有用的
alternative	*n.* 二中择一，可供选择的办法或事物
	adj. 选择性的，二中择一的
upgrade	*n.* 升级
	v. 使升级，提升
replacement	*n.* 代替，置换
encryption	*n.* 加密，编密码
preinstall	*n.* 预安装
enforce	*v.* 执行，坚持，加强
protocol	*n.* 协议
carelessness	*n.* 粗心大意，草率
ignorance	*n.* 无知，不知
innovative	*adj.* 创新的
accessible	*adj.* 易接近的；可到达的；易受影响的；可理解的
estimate	*v. & n.* 估计，估价，评估
mode	*n.* 方式，模式
link	*n. & v.* 链接
respective	*adj.* 分别的，各自的
suppression	*n.* 抑制

option	*n.* 选项，选择权
threatening	*adj.* 胁迫的，危险的
snapshot	*n.* 快照；简单印象
scenario	*n.* 情景，情节
accidental	*adj.* 意外的；非主要的；附属的
	n. 非本质属性，次要方面
overlap	*v.* (与……)重叠
hook	*n.* 钩，吊钩
	v. 钩住
deliberate	*adj.* 深思熟虑的；故意的，预有准备的
bypass	*v.* 绕过，设旁路，迂回
	n. 旁路
authentication	*n.* 证明，鉴定
feature	*n.* 特色
	v. 起重要作用
disabled	*adj.* 无效的，关闭的
	v. 丧失能力
unsecured	*adj.* 不安全的
folder	*n.* 文件夹
Bluetooth	*n.* 蓝牙
barcode	*n.* 条形码
handheld	*n.* 手持式
printer	*n.* 打印机
copier	*n.* 复印机
filter	*n.* 过滤器，筛选
	v. 过滤
pretend	*v.* 假装，装扮
hurdle	*v.* 跳过(栏栅)，克服(障碍)
unencrypted	*adj.* 未加密的
header	*n.* 报头
receiver	*n.* 接收器，收信机
adapter	*n.* 适配器
freeware	*n.* 免费软件
analyzer	*n.* 分析器，分析者
casual	*adj.* 偶然的，不经意的，临时的
unintended	*adj.* 非故意的，无意的
entice	*v.* 诱惑，诱使
sniff	*v.* 嗅探
execute	*v.* 执行，实行

reconnect	*v.*	再连接
disconnect	*v.*	拆开，分离，断开
modem	*n.*	调制解调器
extract	*v.*	提取，析取
vulnerable	*adj.*	易受攻击的
continually	*adv.*	不断地，频繁地
bombard	*v.*	炮轰，轰击
bogus	*adj.*	假的，伪造的
premature	*adj.*	未成熟的，太早的，早熟的
command	*n.* & *v.*	命令
abuse	*n.* & *v.*	滥用
interruption	*n.*	中断，打断
indirectly	*adv.*	间接地
recovery	*n.*	恢复，痊愈，防御
retransmit	*v.*	转播，转发，中继站发送
nonfiltered	*adj.*	没有过滤的
broadcasting	*n.*	广播
reconfiguration	*n.*	重新配置
intelligent	*adj.*	智能的
hub	*n.*	网络集线器，网络中心
reboot	*n.*	重新启动，重新引导
reprogram	*v.*	重新编程，程序重调
restriction	*n.*	限制，约束
isolated	*adj.*	隔离的，孤立的，单独的
backdoor	*n.*	后门
independent	*adj.*	独立的，不受约束的
consensus	*n.*	一致同意，多数人的意见，舆论
robust	*adj.*	健壮的，可靠的
counteract	*v.*	抵消，中和，阻碍
overlay	*n.*	覆盖
guideline	*n.*	方针

Phrases

wireless security	无线安全
wireless network	无线网络
laptop computer	笔记本式计算机
wireless card	无线网卡
move around	走来走去
be prone to	有……的倾向，易于

break into	攻入，侵占
as a result	结果
guard against	提防，预防
latch on to	理解，明白，取得，得到
associate with	联合
at no charge	免费，不收费
wireless access point	无线接入点
be equipped with	装备
plug into	插入，接入
to some extent	在某种程度上，有一点
rely on	依赖，依靠
lure into	引诱（某人或某动物）进入（某处），诱骗（某人）（做某事）
launch on	开始，着手
security measure	安全措施，保密措施
ad hoc network	自组织网络，特定网络
security hole	安全漏洞
turn on	开启，开始
be regarded as	被视作，被认为，被视为
listen in on	偷听，窃听
network traffic	网络流
off the air	不传播
on the air	传播中
directed attack	被控制的攻击，定向攻击
man-in-the-middle attack	中间人攻击
challenge and handshake protocol	挑战和握手协议
de-authentication attack	取消验证攻击
script kiddie	脚本小子
handshake code	握手代码
cracking tool	破解工具
network injection attack	网络注入攻击
intelligent hub	智能集线器
in this manner	以这种方式
check on	检查
captive portal	强制网络门户
hook up	以钩钩住，挂上
end-to-end encryption	端到端加密
as a whole	总体上
state of the art	最新水平
Payment Card Industry Security Standards Council	支付卡行业安全标准委员会

Abbreviations

WEP (Wired Equivalent Privacy)	有线等效加密
WPA (WiFi Protected Access)	无线保护接入，无线保护访问
IEEE (Institute of Electrical and Electronics Engineers)	电气和电子工程师协会
WIPS (Wireless Intrusion Prevention System)	无线入侵防御系统
WIDS (Wireless Intrusion Detection System)	无线入侵检测系统
AP (Access Point)	接入点
LAN (Local Area Network)	局域网
PDA (Personal Digital Assistant)	个人数字助理
IT (Information Technology)	信息技术
MAC (Media Access Control)	介质访问控制，介质接入控制
SOHO (Small Office，Home Office)	家居办公
DoS (Denial of Service)	拒绝服务
EAP (Extensible Authentication Protocol)	扩展认证协议
OSPF (Open Shortest Path First)	开放最短路径优先协议
RIP (Routing Information Protocol)	路由信息协议
HSRP (Hot Standby Router Protocol)	热备份路由协议
VPN (Virtual Private Network)	虚拟个人网络，虚拟私有网络
PCI DSS (Payment Card Industry Data Security Standard)	支付卡行业安全标准

Notes

[1] Wired Equivalent Privacy (WEP) is a security algorithm for IEEE 802.11 wireless networks. Introduced as part of the original 802.11 standard ratified in 1997, its intention was to provide data confidentiality comparable to that of a traditional wired network. WEP, recognizable by its key of 10 or 26 hexadecimal digits (40 or 104 bits), was at one time widely in use and was often the first security choice presented to users by router configuration tools.

[2] WiFi Protected Access (WPA) and WiFi Protected Access II (WPA2) are two security protocols and security certification programs developed by the WiFi Alliance to secure wireless computer networks. The Alliance defined these in response to serious weaknesses researchers had found in the previous system, Wired Equivalent Privacy (WEP).

[3] In computing, a wireless intrusion prevention system (WIPS) is a network device that monitors the radio spectrum for the presence of unauthorized access points (intrusion detection), and can automatically take countermeasures (intrusion prevention).

[4] A virtual private network (VPN) extends a private network across a public network, such as the Internet. It enables users to send and receive data across shared or public networks as if their computing devices were directly connected to the private network. Applications running across the VPN may therefore benefit from the functionality, security, and management of the private network.

[5] Bluetooth is a wireless technology standard for exchanging data over short distances (using

short-wavelength UHF radio waves in the ISM band from 2.4 to 2.485 GHz) from fixed and mobile devices, and building personal area networks (PANs).

[6] MAC spoofing is a technique for changing a factory-assigned Media Access Control (MAC) address of a network interface on a networked device. The MAC address that is hard-coded on a network interface controller (NIC) cannot be changed. However, many drivers allow the MAC address to be changed. Additionally, there are tools which can make an operating system believe that the NIC has the MAC address of a user's choosing. The process of masking an MAC address is known as MAC spoofing. Essentially, MAC spoofing entails changing a computer's identity, for any reason, and it is relatively easy.

[7] In computing, the Challenge-Handshake Authentication Protocol (CHAP) authenticates a user or network host to an authenticating entity. That entity may be, for example, an Internet service provider.

CHAP provides protection against replay attacks by the peer through the use of an incrementally changing identifier and of a variable challenge-value. CHAP requires that both the client and server know the plaintext of the secret, although it is never sent over the network. Thus, CHAP provides better security as compared to Password Authentication Protocol (PAP) which is vulnerable for both these reasons.

[8] In programming and hacking culture, a script kiddie or skiddie (other names include skid or script bunny) is an unskilled individual who uses scripts or programs developed by others to attack computer systems and networks and deface websites. It is generally assumed that script kiddies are juveniles who lack the ability to write sophisticated programs or exploits on their own and that their objective is to try to impress their friends or gain credit in computer-enthusiast communities. However, the term does not relate to the actual age of the participant. The term is generally considered to be pejorative.

[9] Extensible Authentication Protocol, or EAP, is an authentication framework frequently used in wireless networks and point-to-point connections.

[10] Open Shortest Path First (OSPF) is a routing protocol for Internet Protocol (IP) networks. It uses a link state routing (LSR) algorithm and falls into the group of interior routing protocols, operating within a single autonomous system (AS).

[11] The Routing Information Protocol (RIP) is one of the oldest distance-vector routing protocols which employ the hop count as a routing metric. RIP prevents routing loops by implementing limit on the number of hops allowed in a path from source to destination.

[12] A captive portal is a "landing" web page, presented by a Layer 3 brand or Layer 2 Operator and shown to users before they gain broader access to URL or http-based Internet services. Often used to present a landing or log-in page, the portal intercepts observed packets until such time as the user is authorized to launch browser sessions. After being redirected to a web page which may require authentication, payment, acceptance of EULA/acceptable use policies or other valid credentials that the host and user agree to, the user is granted conditional Internet access.

[13] Back Orifice (BO) is a controversial computer program designed for remote system administration. It enables a user to control a computer running the Microsoft Windows operating system from a remote location. The name is a play on words on Microsoft BackOffice Server

software. It can also control multiple computers at the same time using imaging.

Exercises

[Ex. 1] Answer the following questions according to the text.

1. What is wireless security?
2. What are commonly used to enforce wireless security policies?
3. What is accidental association?
4. What are malicious associations?
5. What is the security hole provided by ad hoc networking?
6. When does identity theft (or MAC spoofing) occur?
7. What does a man-in-the-middle attacker entice computers to do?
8. How are man-in-the-middle attacks enhanced?
9. What is the usual reason for performing a DoS attack?
10. What are the three principal ways to secure a wireless network?

[Ex. 2] Translate the following terms or phrases from English into Chinese and vice versa.

1.	cracking tool	1.	
2.	handshake code	2.	
3.	directed attack	3.	
4.	man-in-the-middle attack	4.	
5.	network traffic	5.	
6.	滥用	6.	
7.	适配器	7.	
8.	分析器，分析者	8.	
9.	条形码	9.	
10.	加密，编密码	10.	

[Ex. 3] Translate the following passage into Chinese.

There's no question that apps can make your life easier. Smartphone apps can help you get organized, give you directions when you're lost, find the nearest gas station or ATM, entertain you, help you get in shape and much more. But while you're sharing your life with your apps, how much information about you are those useful tools broadcasting to the rest of the world?

Most people aren't surprised to learn that apps can (and do) collect your personal data. A cursory glance at the Terms of Service (ToS) of any app reveals this information. However, you may not know what kind of data your apps are collecting—or who they're sharing it with.

Nearly every app you use, particularly if it's free, shares your information in some form. In most cases, that personal data is generalized and shared non-specifically with advertisers for a variety of reasons—usually to generate more targeted ads that match your habits, preferences and likely interests.

Other popular apps that may be sharing too much include:

- Angry Birds: Accesses your phone ID, contacts and location data (and shares your location with third parties).

- Pandora: Accesses phone ID, location data and contacts (and shares your contacts).
- TextPlus 4: Accesses and sends phone ID information to ad companies.
- Facebook: Many third-party apps on Facebook are heavy offenders in over sharing your private data.

[Ex. 4] Fill in the blanks with the words given below.

policy	prevent	collecting	update	download
apps	advertisers	shared	location	smartphone

There are thousands of apps that let you track your health, create a personalized diet or fitness plan and help you organize your overall health strategies. In order to work, these apps ask for a lot of personal information, and some of them turn around and sell that information to __1__.

A report released by the FTC in May 2014 looked at a dozen health and fitness apps, and found that these __2__ collectively sent personal data to 76 different third parties. Among the data shared were names and email addresses, __3__, gender, diets, exercise habits and medical symptom searches.

While some apps will disclose how user information is being used upfront, others might not be so straightforward. Here are steps you can take to __4__ your data from being overshared with third parties through.

- Research apps and the companies that distribute them, including user reviews and mentions in industry publications, before you __5__ them.
- Read the entire privacy __6__ and, if you have an Android phone, read the entire "Permissions" screen as well.
- When prompted by the app, opt out of location sharing.
- Check the privacy settings on your __7__ periodically and make sure they're set as high as possible without compromising app functions. For example, map and directions apps like Google require your geolocation.
- Always __8__ your apps when you're alerted to a new version, as these updates often repair "bugs" found in earlier versions.
- Delete any apps you're no longer using from your phone.

Neither smartphones nor Web-based apps are likely to stop __9__ information from users, but you can make sure that information isn't __10__ in ways you aren't comfortable with. Pay attention to the fine print, and keep your personal information personal.

Text B

Mobile Security

Mobile security or mobile phone security has become increasingly important in mobile computing[1]. Of particular concern is the security of personal and business information now stored on smartphones.

Since smartphones collect and compile an increasing amount of sensitive personal and business information every day, they have become preferred targets of attacks.

1. Challenges of Mobile Security

(1) Threats

A smartphone user is exposed to various threats when they use their phone. These threats can disrupt the operation of the smartphone and transmit or modify user data.

There are three prime targets for attackers:

- Data: Smartphones are devices for data management, and may contain sensitive data like credit card numbers, authentication information, private information, activity logs (calendar, call logs),etc.
- Identity: Smartphones are highly customizable, so the device or its contents can easily be associated with a specific person. For example, every mobile device can transmit information related to the owner of the mobile phone contract, and an attacker may want to steal the identity of the owner of a smartphone to commit other offenses.
- Availability: Attacking a smartphone can limit access to it and deprive the owner of its use.

The source of these attacks are the same actors found in the non-mobile computing space:

- Professionals, whether commercial or military, who focus on the three targets mentioned above. They steal sensitive data from the general public, as well as undertake industrial espionage. They will also use the identity of those attacked to achieve other attacks.
- Thieves who want to gain income through data or identities they have stolen. The thieves will attack many people to increase their potential income.
- Black hat hackers who specifically attack availability. Their goal is to develop viruses, and cause damage to the device. In some cases, hackers have an interest in stealing data on devices.
- Grey hat[2] hackers who reveal vulnerabilities. Their goal is to expose vulnerabilities of the device. Grey hat hackers do not intend to damage the device or steal data.

(2) Consequences

When a smartphone is attacked by an attacker, the attacker can attempt several things:

- The attacker can manipulate the smartphone as a zombie[3] machine, that is to say, a machine with which the attacker can communicate and send commands which will be used to send unsolicited messages (spam) via SMS or e-mail.
- The attacker can easily force the smartphone to make phone calls. For example, one can use the API PhoneMakeCall by Microsoft, which collects telephone numbers from any source such as yellow pages, and then call them. But the attacker can also use this method to call paid services, resulting in a charge to the owner of the smartphone. It is also very dangerous because the smartphone could call emergency services and thus disrupt those services.
- A compromised smartphone can record conversations between the user and others and send them to a third party. This can cause user privacy and industrial security problems.

- An attacker can also steal a user's identity, usurp their identity (with a copy of the user's SIM[4] card or even the telephone itself), and thus impersonate the owner. This raises security concerns in countries where smartphones can be used to place orders, view bank accounts or are used as an identity card.
- The attacker can reduce the utility of the smartphone by discharging the battery. For example, they can launch an application that will run continuously on the smartphone processor, requiring a lot of energy and draining the battery. One factor that distinguishes mobile computing from traditional desktop PCs is their limited performance. Frank Stajano and Ross Anderson first described this form of attack, calling it an attack of "battery exhaustion" or "sleep deprivation torture".
- The attacker can prevent the operation and/or starting of the smartphone by making it unusable. This attack can either delete the boot scripts, resulting in a phone without a functioning OS[5], or modify certain files to make it unusable (e.g. a script that launches at startup that forces the smartphone to restart) or even embed a startup application that would empty the battery.
- The attacker can remove the personal (photos, music, videos, etc.) or professional data (contacts, calendars, notes) of the user.

2. Attacks Based on Communication

Some attacks derive from flaws in the management of SMS and MMS.

Some mobile phone models have problems in managing binary SMS messages. It is possible, by sending an ill-formed block, to cause the phone to restart, leading to denial of service attacks. If a user with a Siemens S55 received a text message containing a Chinese character, it would lead to a denial of service. In another case, while the standard requires that the maximum size of a Nokia Mail address is 32 characters, some Nokia phones did not verify this standard, so if a user enters an e-mail address over 32 characters, that will lead to complete dysfunction of the e-mail handler and put it out of commission. This attack is called "curse of silence". A study on the safety of the SMS infrastructure revealed that SMS messages sent from the Internet can be used to perform a Distributed Denial of Service (DDoS)[6] attack against the mobile telecommunications infrastructure of a big city. The attack exploits the delays in the delivery of messages to overload the network.

Another potential attack could begin with a phone that sends an MMS to other phones, with an attachment. This attachment is infected with a virus. Upon receipt of the MMS, the user can choose to open the attachment. If it is opened, the phone is infected, and the virus sends an MMS with an infected attachment to all the contacts in the address book. There is a real-world example of this attack: The virus Commwarrior uses the address book and sends MMS messages including an infected file to recipients. A user installs the software, as received via MMS message. Then, the virus began to send messages to recipients taken from the address book.

3. Attacks Based on Communication Networks

(1) Attacks Based on WiFi

An attacker can try to eavesdrop on WiFi communications to derive information (e.g. username, password). This type of attack is not unique to smartphones, but they are very vulnerable to these attacks because very often the WiFi is the only means of communication they have to access the Internet. The security of wireless networks (WLAN) is thus an important subject. Initially wireless networks were secured by WEP keys. The weakness of WEP is a short encryption key which is the same for all connected clients. In addition, several reductions in the search space of the keys have been found by researchers. Now, most wireless networks are protected by the WPA security protocol. WPA is based on the "Temporal Key Integrity Protocol (TKIP)" which was designed to allow migration from WEP to WPA on the equipment already deployed. The major improvements in security are the dynamic encryption keys. For small networks, the WPA is a "pre-shared key[7]" which is based on a shared key. Encryption can be vulnerable if the length of the shared key is short. With limited opportunities for input (i.e. only the numeric keypad) mobile phone users might define short encryption keys that contain only numbers. This increases the likelihood that an attacker succeeds with a brute force attack. The successor to WPA, called WPA2, is supposed to be safe enough to withstand a brute force attack.

As with GSM, if the attacker succeeds in breaking the identification key, it will be possible to attack not only the phone but also the entire network it is connected to.

Many smartphones for wireless LANs remember they are already connected, and this mechanism prevents the user from having to re-identify with each connection. However, an attacker could create a WiFi access point twin with the same parameters and characteristics as the real network (see Figure 2-1). Using the fact that some smartphones remember the networks, they could confuse the two networks and connect to the network of the attacker who can intercept data if it does not transmit its data in encrypted form.

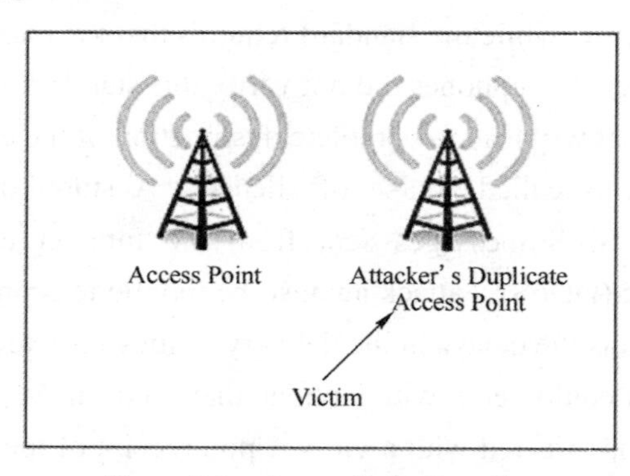

Figure 2-1 Access Point Spoofing

An attacker can try to eavesdrop on WiFi communications to derive information (e.g. username, password). This type of attack is not unique to smartphones, but they are very vulnerable to these attacks because very often the WiFi is the only means of communication they have to access the internet. The security of Wireless Networks (WLAN) is thus an important subject. Initially wireless networks were secured by WEP keys. The weakness of WEP is a short encryption key which is the same for all connected clients. In addition, several reductions in the search space of the keys have been found by researchers. Now, most wireless networks are protected by the WPA security protocol.

WPA is based on the "Temporal Key Integrity Protocol (TKIP)" which was designed to allow migration from WEP to WPA on the equipment already deployed. The major improvements in security are the dynamic encryption keys. For small networks, the WPA is a "pre-shared key[7]" which is based on a shared key. Encryption can be vulnerable if the length of the shared key is short. With limited opportunities for input (i.e. only the numeric keypad) mobile phone users might define short encryption keys that contain only numbers. This increases the likelihood that an attacker succeeds with a brute force attack. The successor to WPA, called WPA2, is supposed to be safe enough to withstand a brute force attack.

As with GSM, if the attacker succeeds in breaking the identification key, it will be possible to attack not only the phone but also the entire network it is connected to.

Many smartphones for wireless LANs remember they are already connected, and this mechanism prevents the user from having to re-identify with each connection. However, an attacker could create a WiFi access point with the same parameters and characteristics as the real network (see Figure 2-1). Using the fact that some smartphones remember the networks, they could confuse the two networks and connect to the network of the attacker who can intercept data if it does not transmit its data in encrypted form.

Lasco is a worm that initially infects a remote device using the SIS (Software Installation Script) file format. SIS file format is a script file that can be executed by the system without user interaction. The smartphone thus believes the file to come from a trusted source and downloads it, infecting the machine.

(2) Principle of Bluetooth-Based Attacks

Security issues related to Bluetooth on mobile devices have been studied and have shown numerous problems on different phones. One vulnerability is that unregistered services do not require authentication, and vulnerable applications have a virtual serial port used to control the phone. An attacker only needs to connect to the port to take full control of the device. Another example: A phone must be within reach and Bluetooth in discoverable mode. The attacker sends a file via Bluetooth. If the recipient accepts, a virus is transmitted. For example: Cabir is a worm that spreads via Bluetooth connection. The worm searches for nearby phones with Bluetooth in discoverable mode and sends itself to the target device. The user must accept the incoming file and install the program. After installing, the worm infects the machine.

New Words

smartphone	*n.*	智能电话
preferred	*adj.*	首选的
target	*n.*	目标，对象
disrupt	*v.*	使中断，破坏
modify	*v.*	更改，修改
log	*n.*	日志
customizable	*n.*	可定制化，可用户化

commit	*v.*	犯(错误)，干(坏事)，提交
deprive	*v.*	剥夺，使丧失
undertake	*v.*	承担，担任，许诺，保证
espionage	*n.*	间谍，侦探
reveal	*v.*	展现，显示，揭示，暴露
zombie	*n.*	僵尸
spam	*n.*	垃圾邮件
conversation	*n.*	会话，交谈
usurp	*v.*	篡夺，侵占
discharge	*v.* & *n.*	卸载，放电
processor	*n.*	处理器
drain	*v.*	耗尽
unusable	*adj.*	不能用的
delete	*v.*	删除
boot	*v.*	导入，引导
restart	*v.*	重新启动
embed	*v.*	使插入，使嵌入，嵌入
dysfunction	*n.*	功能紊乱，机能异常
delivery	*n.*	发送，传输
overload	*v.*	使超载，超过负荷
	n.	超载，过载
attachment	*n.*	附件，附加装置
eavesdrop	*v.*	偷听
migration	*n.*	移植
improvement	*n.*	改进，进步
input	*n.* & *v.*	输入
likelihood	*n.*	可能，可能性
withstand	*v.*	抵挡，经受住
re-identify	*v.*	重新识别，重新鉴别
parameter	*n.*	参数，参量
characteristic	*n.*	特性，特征

Phrases

mobile phone security	手机安全，移动电话安全
mobile computing	移动计算
be exposed to	遭受，暴露于……
credit card number	信用卡号
industrial espionage	工业间谍
potential income	潜在收入

cause damage to	损害，破坏
have an interest in…	对……有兴趣，对……关心
yellow page	黄页
emergency service	应急服务
third party	第三方
identity card	身份证
battery exhaustion	电池耗尽
out of commission	不能使用
curse of silence	沉默的诅咒
be infected with	感染，沾染上
address book	地址簿
numeric keypad	数字小键盘
succeed with…	在……上获得成功
brute force	强力
serial port	串行端口
discoverable mode	可发现模式

Abbreviations

SMS (Short Message Service)	短信，手机短信服务
API (Application Programming Interface)	应用程序编程接口
SIM (Subscriber Identification Module)	客户识别模块
MMS (Multimedia Messaging Service)	彩信
WLAN (Wireless Local Area Network)	无线局域网
TKIP (Temporal Key Integrity Protocol)	临时密钥完整性协议
GSM (Global System for Mobile Communication)	全球移动通信系统
SIS (Software Installation Script)	软件安装脚本

Notes

[1] Mobile computing is human-computer interaction by which a computer is expected to be transported during normal usage, which allows for transmission of data, voice and video. Mobile computing involves mobile communication, mobile hardware, and mobile software. Communication issues include ad hoc networks and infrastructure networks as well as communication properties, protocols, data formats and concrete technologies. Hardware includes mobile devices or device components. Mobile software deals with the characteristics and requirements of mobile applications.

[2] The term "grey hat" refers to a computer hacker or computer security expert who may sometimes violate laws or typical ethical standards, but does not have the malicious intent typical of a black hat hacker.

[3] In computer science, a zombie is a computer connected to the Internet that has been compromised by a hacker, computer virus or trojan horse program and can be used to perform malicious tasks of one sort or another under remote direction. Botnets of zombie computers are often

used to spread e-mail spam and launch Denial-of-Service attacks (DoS attacks). Most owners of "zombie" computers are unaware that their system is being used in this way. Because the owner tends to be unaware, these computers are metaphorically compared to fictional zombies.

[4] A subscriber identity module or Subscriber Identification Module (SIM) is an integrated circuit that is intended to securely store the International Mobile Subscriber Identity (IMSI) number and its related key, which are used to identify and authenticate subscribers on mobile telephony devices (such as mobile phones and computers). It is also possible to store contact information on many SIM cards. SIM cards are always used on GSM phones; for CDMA phones, they are only needed for newer LTE-capable handsets. SIM cards can also be used in satellite phones.

[5] A mobile operating system (or mobile OS) is an operating system for smartphones, tablets, Personal Digital Assistants (PDAs), or other mobile devices. While computers such as typical laptops are mobile, the operating systems usually used on them are not considered mobile ones, as they were originally designed for desktop computers that historically did not have or need specific mobile features. This distinction is becoming blurred in some newer operating systems that are hybrids made for both uses. So-called mobile operating systems, or even only smartphones running them, now represent most (web) use (on weekends and averaged for whole weeks).

[6] DDoS (Distributed Denial of Service) attacks are malicious attempts to render a web site or web application unavailable to users by overwhelming the site with an enormous amount of traffic, causing the site to crash or operate very slowly.

[7] In cryptography, a Pre-Shared Key (PSK) is a shared secret which was previously shared between the two parties using some secure channel before it needs to be used.

To build a key from shared secret, the key derivation function is typically used. Such systems almost always use symmetric key cryptographic algorithms. The term PSK is used in WiFi encryption such as Wired Equivalent Privacy (WEP), WiFi Protected Access (WPA), where the method is called WPA-PSK or WPA2-PSK, and also in the Extensible Authentication Protocol (EAP), where it is known as EAP-PSK. In all these cases, both the wireless Access Points (AP) and all clients share the same key.

Exercises

[Ex. 5] Answer the following questions according to the text.

1. How many prime targets are there for attackers? What are they?
2. What are the source of these attacks?
3. What is the goal of black hat hackers?
4. What is the goal of grey hat hackers?
5. What is a zombie machine?
6. What is a compromised smartphone?
7. What is the weakness of WEP?
8. What will happen if the attacker succeeds in breaking the identification key as with GSM?
9. What could attackers do by using the fact that some smartphones remember the networks?

10. What is Lasco?

Reading Material

Solutions to Mobile Security Threats

Threats to mobile security are becoming varied[1] and stronger. Managing mobile security is a big challenge for a number of reasons. Traditional IT security and mobile security are different propositions to a great extent[2]. That is why the approach to mobile security needs to be different. A number of strategies are being implemented, including dual OS, remote wiping[3], secure browsing and app lifecycle management. While enterprises are working on improving security practices, awareness needs to grow at the individual level as well.

1. Implementing Secure OS Architecture

Implementation[4] of a secure OS architecture has already begun with iPhones and the latest Samsung Galaxy Android smartphones implementing the feature. The iPhone and the Samsung Galaxy smartphones have two OSs: One OS is known as the application OS and the other is a smaller and more secure OS. The application OS is where smartphone users download and run their apps, while the second OS is used to handle keychain and cryptographic functions as well as other high-security tasks.

According to a white paper on Apple's secure mobile OS, "The Secure Enclave is a coprocessor[5] fabricated in the Apple A7 or later A-series processor. It utilizes its own secure boot and personalized software update separate from[6] the application processor."

So, the secure OS communicates with the application OS over a shared, and probably unencrypted, memory space[7] and a single mailbox[8]. The application OS is not allowed to access the the main memory of the secure OS. Certain devices such as the touch ID sensor communicate with the secure OS over an encrypted channel[9]. Samsung Galaxy smartphones use the TrustZone-based Integrity Measurement Architecture (TIMA) to verify the Android OS' integrity.

Since a large number of financial transactions happen over mobile devices, the dual OS system could be extremely handy[10]. For example, in the case of [11]a credit card transaction, the secure OS

---

1 varied adj. 多变的，各种各样的，各不相同的，形形色色的

2 to a great extent 在很大程度上，非常

3 remote wiping 远程清除

4 implementation *n.* 执行

5 coprocessor *n.* 协处理器

6 separate from 分离，分开

7 memory space 存储量，存储空间

8 mailbox *n.* 邮箱

9 channel *n.* 信道，频道

10 handy *adj.* 手边的，唾手可得的，便利的，容易取得的

11 in the case of 在……的情况

will handle and pass the credit card data in an encrypted format. The application OS cannot even decrypt[1] it.

2. Introducing Encryption and Authentication

Encryption and authentication have been implemented in smartphones to some degree already, but these steps are not enough. Recently, different concepts have been implemented to make encryption and authentication more robust. One such concept is containers[2]. Simply put, containers are third-party applications that isolate[3] and secure a certain portion of a smartphone's storage. It is like a high-security zone. The goal is to prevent intruders, malware, system resources or other applications from accessing the application or its sensitive data.

Containers are available on all popular mobile OSs: Android, Windows, iOS and BlackBerry. Samsung offers Knox, and VMware provides containers for Android from the Horizon Mobile technology. Containers are available both for personal use and at the enterprise level.

Another way of encrypting mobile devices is to introduce compulsory encryption. Google is doing that with Android Marshmallow, and all devices that run Marshmallow are required to make use of full-disk encryption out of the box. Although earlier Android OS versions allowed one to enable encryption, i.e. since Android 3.0, the option had two limitations. One was that it was an optional task (only Nexus devices were shipped with encryption already enabled) so users did not usually enable it, and the other was that enabling encryption was a bit too technical for many typical users.

3. Implementing Network Security and Secure Browsing

From the mobile device user's point of view, there are a number of ways to browse securely:

- Do not modify the default browser settings in Android, iOS or Windows devices because the default settings are already providing good security.
- Do not log into unencrypted public wireless networks. People with bad intentions can also log into them. Sometimes, malicious people can set up an open network and set a trap for unsuspecting[4] users.
- Try to use wireless networks that are secured. Such networks need a password or other authentication to allow access.
- Whenever you access a website where you are going to share personal or confidential[5] information, such as your bank account details, make sure that the URL begins with HTTPS. This means that all data transmitted through this website is encrypted.

While secure browsing is required, it is at best to secure mobile devices. The foundation[6] is always network security. Mobile device security should begin with a multi-layered approach such as

---

1 decrypt *v.* 解密

2 container *n.* 容器

3 isolate *v.* 使隔离，使孤立

4 unsuspecting *adj.* 不怀疑的，无疑虑的

5 confidential *adj.* 秘密的，机密的

6 foundation *n.* 基础，根本

VPN, IPS[1], firewall and application control. Next-generation firewalls and unified threat management help IT administrators to monitor the flow of data and the behavior of users and devices while connected to a network.

4. Implementing Remote Wipe

Remote wipe is the practice of wiping out[2] data from a mobile device via a remote location. This is done to make sure that confidential data does not fall into[3] unauthorized hands. Normally, remote wipe is used in the following situations:

- The device is lost or stolen.
- The device is with an employee who is no longer with the organization.
- The device contains malware which can access confidential data.

However, since mobile device owners do not want anyone or anything else to access their personal devices, remote wiping may face a limitation. Owners are also rather lethargic when it comes to security.

To overcome these problems, enterprises could create containers in mobile devices which will contain only confidential data. Remote wiping will be exercised only on the container and not on data outside the container. Employees need to feel confident[4] that remote wiping is not going to affect their personal data. Enterprises can track the usage of the mobile device. If the device is not being used for a long time, chances are that it has been lost or stolen. In such a case, the remote wipe should be immediately deployed so that all confidential data is wiped out.

5. App Lifecycle Management and Data Sharing

Application Lifecycle Management[5] (ALM) is the practice of supervising[6] a software application from its primary and initial planning through to the time when the software is retired. The practice also means that changes to the application during the entire lifecycle are documented and the changes can be tracked. Obviously, security of the apps is of primary consideration before any app is made commercially available. It is extremely important to document and track how the security features of the app have evolved over time based on experience and feedback[7] and how it has solved the problems of mobile device security. Depending on how well the security elements are incorporated in the apps, the retirement[8] time for an app or its version is determined.

6. Conclusion

While remote wiping and secure browsing are good practices to follow, the most critical practices for ensuring[9] mobile security are network security, OS architecture security and app life cycle management. These are the foundation pillars[10] based on which a mobile device can be

---

1 IPS (Intrusion Prevention System)　入侵防御系统
2 wipe out　消灭
3 fall into　落入
4 confident　*adj.* 自信的，确信的
5 应用程序生命周期管理
6 supervise　*v.* 监督，管理，指导
7 feedback　*n.* 反馈
8 retirement　*n.* 退休，退出
9 ensure　*v.* 确保，保证
10 pillar　*n.* 柱子，栋梁

judged as secure or relatively unsecure. Over time, these practices must be enhanced as the usage of mobile devices for financial and enterprise transactions grow exponentially[1]. Naturally, that will involve a lot of data being transmitted. The dual OS system followed by Apple seems to be a good case study of how to internally strengthen[2] a mobile device and can be a model for future developments.

参 考 译 文

无 线 安 全

无线安全是措防止对使用无线网络的计算机进行未授权访问或损坏。最常见的无线安全类型是有线等效保密（WEP）和 WiFi 保护访问（WPA）。众所周知，WEP 是一个弱安全标准。它使用的密码通常可以在几分钟内用普通的笔记本式计算机和广泛使用的软件工具破解。WEP 是 1999 年就使用的旧 IEEE 802.11 标准，2003 年被 WPA 淘汰。WPA 是提高 WEP 安全性的快速替代方法。当前标准是 WPA2。某些硬件不升级、不更换固件就不能支持 WPA2。WPA2 使用一个用 256 位密钥加密网络的加密设备；更长的密钥长度提高了 WEP 的安全性。

许多笔记本式计算机预装了无线网卡，这就让人们在移动时可以入网。然而，无线网络容易出现一些安全问题。黑客已经发现无线网络相对容易攻入，甚至可以使用无线技术侵入有线网络。因此，企业制定有效的无线安全策略以防止未经授权访问重要资源，这一点非常重要。无线入侵防御系统（WIPS）或无线入侵检测系统（WIDS）通常用于实施无线安全策略。

无线服务越受欢迎，使用无线技术的用户面临的风险就越大。当无线技术首次引入时，存在的危险相对较少。那是因为黑客还没有时间掌握新技术，而且在工作场所中无线网络并不常见。然而，当前无线协议和加密方法有许多安全风险，用户和公司 IT 层对此并不知晓也不够重视。黑客进入无线网络的方法日益复杂并不断创新。在网上可以免费获得易用的基于 Windows 或 Linux 的工具，这让黑客访问无线网络更容易。

一些没有安装无线接入点的组织觉得他们没有无线安全问题。当无线笔记本式计算机插入公司网络时，在没有无线网络的组织中可能出现问题。黑客可以坐在停车场，通过笔记本式计算机和/或其他设备收集信息，甚至通过这个配有无线网卡的笔记本式计算机访问有线网络。

1. 未经授权的访问模式

未经授权访问链接、功能和数据的模式与各实体使用程序代码一样，是多变的。这种威胁没有一个完整的模式。在某种程度上，预防依赖于已知的攻击模式和方法，以及用于抑制所应用的方法的相关方法。然而，每种新的操作模式都将产生新的威胁形式。因此，预防需要持续地改进。下面所描述的攻击模式只是一些典型方法和使用情景。

（1）意外关联

多种不同的方法和意图都能突破公司网络的安全边界。其中之一被称为“意外关联”。

---

1 exponential *adj.* 指数的，幂数的

2 strengthen *v.* 加强，巩固

当用户打开计算机并且从相邻公司的重叠网络进入无线接入点时，用户甚至可能不知道这已经发生。然而，这是一个安全漏洞，因为专有公司信息已经暴露，现在可能建立了从一个公司到另一个公司的链接。如果笔记本式计算机也连接到有线网络，则更是如此。

意外关联是一种被称为“误关联”的无线漏洞。误关联可能是偶然的、故意的（例如，绕过公司防火墙），也可能有意诱使无线客户连接到攻击者的AP。

（2）恶意关联

“恶意关联”是指攻击者主动地用其笔记本式计算机而不是公司接入点（AP）把无线设备连接到公司网络。这些类型的笔记本式计算机被称为“软AP”，当网络犯罪分子运行一些使他/她的无线网卡看起来像合法接入点的软件时就创建了这类连接。一旦窃贼获得访问权限，他/她便可以窃取密码，在有线网络上发起攻击或植入木马。由于无线网络在第2层级运行，第3层保护（例如网络身份验证和虚拟专用网络（VPN））不能提供任何屏障。无线802.1x身份验证的确能提供一些保护，但仍然容易受到黑客攻击。这种类型的攻击背后的想法可能不是攻入VPN或破坏其他安全措施，很可能犯罪分子只是试图在第2层级接管客户。

（3）自组织网络

自组织网络可能会带来安全威胁。自组织网络被定义为对等网络，在没有接入点的无线计算机之间建立连接。虽然此类网络通常几乎没有保护，但是可以使用加密方法来提供安全性。

自组织网络提供的安全漏洞不是自组织网络本身。它提供了连接到其他网络的桥梁，通常用在企业环境中。不幸的是，这些企业都使用大多数Microsoft Windows版本中的默认设置。除非明确禁用，否则此功能将打开。因此，用户甚至可能不知道他们操作的计算机运行在不安全的自组织网络中。如果它们同时也在使用有线或无线基础设施网络，那么它们提供了把不安全的自组织网络连接到安全组织网络的网桥。网桥有两种形式：一种是直接桥接，它要求用户实际上在两个连接之间配置网桥，因此除非明确期望，否则不可能连接。另一种是间接桥接，共享用户计算机上的资源。间接网桥可以将从用户计算机共享的私有数据暴露给LAN连接，如共享文件夹或专用网络附属存储，而不能把认证或私有连接与未认证的自组织网络加以区分。

（4）非传统网络

像个人网络蓝牙设备这样的非传统网络受到攻击便不安全，应被视为安全风险。甚至条码阅读器、手持PDA、无线打印机和复印机也应该保证其安全性。这些非传统网络很容易被那些只专注于笔记本式计算机和接入点安全的IT人员忽视。

（5）身份盗窃（MAC欺骗）

当黑客能够窃听网络流并识别具有网络权限的计算机的MAC地址时，就会发生身份窃取（或MAC欺骗）。大多数无线系统允许某种类型的MAC过滤，仅允许具有特定MAC ID的授权计算机能够访问和使用网络。然而，有些程序具有网络“嗅探”能力。黑客可以把这些程序与其他软件结合使用，这就能让计算机冒充黑客想要的任何MAC地址，黑客很容易绕过MAC过滤。

MAC过滤仅对小型住宅（SOHO）网络有效，因为它仅在无线设备“关闭”时提供保护。任何“开放”的802.11设备都随意发送其802.11报头中其未加密的MAC地址，并且不需要特殊的设备或软件来检测它。任何具有 802.11 接收器（笔记本式计算机和无线适配器）和免费软件无线分组分析器的人都可以获得任何正在发送的 802.11 范围内的 MAC 地址。在组织环境中，大多数无线设备在整个工作期间都是“开放的”，MAC 过滤仅提供虚假的安全感，因为它仅防止对组织基础设施的“临时”或无意的连接，而不能阻止定向的攻击。

（6）中间人攻击

中间人攻击者诱使计算机登录到设置为软 AP 的计算机上。一旦完成，黑客可通过另一无线网卡连接到真实接入点，该无线网卡通过透明黑客计算机向实际网络提供稳定的流量。然后，黑客可以嗅探流量。一类中间人攻击依赖于挑战和握手协议中的安全故障来执行“去认证攻击”。这种攻击迫使 AP 连接的计算机断开其连接并且与黑客的软 AP 重新连接（断开用户与调制解调器的连接，因此用户必须使用他们可以从事件记录中提取的密码再次连接）。中间人攻击通过 LANjack 和 AirJack 等软件得到增强，这些软件可以自动完成这个过程的多个步骤，这意味着以前需要有一些技术的人才能做的工作现在“脚本小子”就能完成。热点特别容易受到各种攻击，因为在这些网络上几乎没有安全性。

（7）拒绝服务（DoS）

当攻击者用伪请求、未成功的连接消息、失败消息和/或其他命令连续地轰击目标 AP 或网络时，就会发生拒绝服务攻击。这会让合法用户无法上网，甚至可能导致网络崩溃。这些攻击依赖于诸如可扩展认证协议（EAP）等协议的滥用。

DoS 攻击本身并不能将组织数据暴露给恶意攻击者，因为网络的中断阻止了数据的流动，并且实际上通过阻止数据的传输间接地保护了数据。执行 DoS 攻击的常见原因是观察无线网络的恢复，在这期间，所有设备重新传输所有初始握手代码，这就为恶意攻击者提供了记录这些代码、使用各种破解工具分析安全弱点，并用其访问未经授权系统的机会。这在诸如 WEP 之类的弱加密系统上最有效。在这些系统中，有许多可用的工具。基于在网络恢复期间捕获的安全密钥“模型”，这些工具能发动“可能接受的”安全密钥的字典式攻击。

（8）网络注入

在网络注入攻击中，黑客可以利用暴露给未过滤网络流的接入点，特别是广播网络流[如“生成树”（802.1D）、OSPF、RIP 和 HSRP]。黑客注入了影响路由器、交换机和智能集线器的虚假网络重新配置命令。这就可以让整个网络崩溃，并且需要重新启动甚至对所有智能网络设备重新编程。

2. 无线入侵防御概念

有三种主要方法来保护无线网络：

1）对于封闭网络（如家庭用户和组织），最常见的方法是在接入点配置访问限制。这些限制可能包括加密和检查 MAC 地址。无线入侵防御系统可以对此网络模型中的无线局域网提供安全保护。

2）对于商业供应商、热点和大型组织，通常首选解决方案是具有开放且未加密但完全隔离的无线网络。起初用户不能访问因特网或任何本地网络资源。商业供应商通常将所有网络流量转发到提供付款和/或授权的强制门户。另一个解决方案是要求用户使用 VPN 安全地连接到特权网络。

3）无线网络不如有线网络安全。在许多办公室中，入侵者可以容易地访问并轻而易举地将它们自己的计算机连接到无线网络，以获得对网络的访问，并且远程入侵者也经常可能通过如 Back Orifice 这样的后门获得对网络的访问。一个常用的解决方案是执行端到端加密，在所有公众不可用的资源上使用独立认证。

没有现成系统能防止无线通信的欺骗性使用，也不能保护通过无线通信的计算机和其他实体的数据和功能。然而，有一个系统，根据对最新的共识，对采取的措施进行整体性合格评审。合格性系统是 ISO / IEC 15408 中规定的国际共识。

无线入侵防御系统（WIPS）是一个概念，它是抵御无线安全风险最有效的方法。然而，这样的 WIPS 不是作为软件包来实现的现成设计解决方案。WIPS 通常被认为覆盖了现有无线 LAN 基础设施，尽管它可以在组织内独立部署以实现无线策略。WIPS 对无线安全非常重要，早在 2009 年 7 月，支付卡行业安全标准委员会就发布了针对 PCI DSS 的无线指南，建议大型组织使用 WIPS 自动化地执行无线扫描和保护。

Unit 3

Text A

Information Security

Information security, sometimes shortened to InfoSec, is the practice of preventing unauthorized access, use, disclosure, disruption, modification, inspection, recording or destruction of information. It is a general term that can be used regardless of the form the data may take (e.g. electronic, physical).

扫一扫，听课文

1. Overview

(1) IT Security

Sometimes referred to as computer security, information technology security is information security applied to technology (most often some form of computer system). It is worthwhile to note that a computer does not necessarily mean a home desktop. A computer is any device with a processor and some memory. Such devices can range from non-networked standalone devices, as simple as calculators, to networked mobile computing devices such as smartphones and tablet computers. IT security specialists are almost always found in any major enterprise/establishment due to the nature and value of the data within larger businesses. They are responsible for keeping all of the technology within the company secure from malicious cyber attacks that often attempt to breach into critical private information or gain control of the internal systems.

The act of providing trust of the information, that is the Confidentiality, Integrity and Availability (CIA) of the information, are not violated. For example, ensuring that data is not lost when critical issues arise. These issues include, but are not limited to, natural disasters, computer/server malfunction or physical theft. Since most information is stored on computers in our modern era, information assurance is typically dealt with by IT security specialists. A common method of providing information assurance is to have an off-site backup of the data in case one of the mentioned issues arise.

(2) Threats

Information security threats come in many different forms. Some of the most common threats today are software attacks, theft of intellectual property, identity theft, theft of information, sabotage, and information extortion. Most people have experienced software attacks of some sort. Viruses, worms, phishing[1] attacks, and Trojan horses are a few common examples of software attacks. The theft of intellectual property has also been an extensive issue for many businesses in the IT field. Identity theft is the attempt to act as someone else usually to obtain that person's personal information or to take advantage of their access to vital information. Theft of information is becoming more prevalent today due to the fact that most devices today are mobile. Sabotage usually consists of the destruction of an organization's website in an attempt to cause loss of confidence on the part of its customers. Information extortion consists of theft of a company's property or information as an attempt to receive a payment in exchange for returning the information or property back to its owner, as with ransomware[2]. There are many ways to help protect yourself from some of these attacks but one of the most functional precautions is user carefulness.

Governments, military, corporations, financial institutions, hospitals and private businesses amass a great deal of confidential information about their employees, customers, products, research and financial status. Most of this information is now collected, processed and stored on electronic computers and transmitted across networks to other computers.

Should confidential information about a business' customers or finances or new product line fall into the hands of a competitor or a black-hat hacker[3], a business and its customers could suffer widespread, irreparable financial loss, as well as damage to the company's reputation. From a business perspective, information security must be balanced against cost; the Gordon-Loeb Model[4] provides a mathematical economic approach for addressing this concern.

For the individual, information security has a significant effect on privacy, which is viewed very differently in different cultures.

The field of information security has grown and evolved significantly in recent years. It offers many areas for specialization, including securing networks and allied infrastructure, securing applications and databases, security testing, information systems auditing, business continuity planning and digital forensics[5].

Possible responses to a security threat or risk are:

- Reduce/mitigate—implement safeguards and countermeasures to eliminate vulnerabilities or block threats.
- Assign/transfer—place the cost of the threat onto another entity or organization such as purchasing insurance or outsourcing.
- Accept—evaluate if cost of countermeasure outweighs the possible cost of loss due to threat.
- Ignore/reject—not a valid or prudent due-care response.

2. Definitions

Information Security Attributes: or qualities, i.e., Confidentiality, Integrity and Availability (CIA). Information Systems are composed in three main portions, hardware, software and communications which meet the standards of information security industry. The mechanisms of protection and prevention has three levels or layers: physical, personal and organizational. Essentially, procedures or policies are implemented to ensure information security within the organizations.

3. Basic Principles

(1) Confidentiality

In information security, confidentiality is the property that information is not made available or disclosed to unauthorized individuals or entities.

(2) Integrity

In information security, data integrity means maintaining and assuring the accuracy and completeness of data over its entire life-cycle. This means that data cannot be modified in an unauthorized or undetected manner. This is not the same thing as referential integrity[6] in databases, although it can be viewed as a special case of consistency as understood in the classic ACID model of transaction processing[7]. Information security systems typically provide message integrity in addition to data confidentiality.

(3) Availability

For any information system to serve its purpose, the information must be available when it is needed. This means that the computing systems used to store and process the information, the security controls used to protect it, and the communication channels used to access it must be functioning correctly. High availability[8] systems aim to remain available at all times, preventing service disruptions due to power outages, hardware failures, and system upgrades. Ensuring availability also involves preventing denial-of-service attacks, such as a flood of incoming messages to the target system essentially forcing it to shut down.

(4) Non-repudiation

In law, non-repudiation implies one's intention to fulfill their obligations to a contract. It also implies that one party of a transaction cannot deny having received a transaction nor can the other party deny having sent a transaction. Note: This is also regarded as part of Integrity.

It is important to note that while technology such as cryptographic systems can assist in non-repudiation efforts, the concept is at its core a legal concept transcending the realm of technology. It is not, for instance, sufficient to show that the message matches a digital signature signed with the sender's private key, and thus only the sender could have sent the message and nobody else could have altered it in transit. The alleged sender could in return demonstrate that the digital signature algorithm is vulnerable or flawed, or allege or prove that his signing key has been compromised. The fault for these violations may or may not lie with the sender himself, and such assertions may or may not relieve the sender of liability, but the assertion would invalidate the claim that the signature necessarily proves authenticity and integrity and thus prevents repudiation.

New Words

disclosure	*n.*	公开，泄露，揭露，披露
disruption	*n.*	中断，破坏
modification	*n.*	更改，修改，修正
inspection	*n.*	检查，视察
destruction	*n.*	破坏，毁灭

worthwhile	*adj.*	值得做的；值得花时间的；合算的
desktop	*n.*	桌面
memory	*n.*	存储器，内存
calculator	*n.*	计算器
specialist	*n.*	专家
establishment	*n.*	公司
violate	*v.*	违犯，冒犯，干扰，违反，妨碍
issue	*n.*	问题；（报刊的）期，号
	v.	发行，发布
malfunction	*n.*	故障
assurance	*n.*	确信，保证
backup	*n.*	备份
	v.	做备份
threat	*n.*	威胁
sabotage	*n.*	破坏
	v.	从事破坏活动；对……采取破坏行动，妨害，破坏
extortion	*n.*	勒索，敲诈
sort	*n.*	种类，类别
	v.	分类，拣选
vital	*adj.*	重大的，至关重要的
prevalent	*adj.*	普遍的，流行的
capacity	*n.*	容量
confidence	*n.*	信心
ransomware	*n.*	敲诈勒索软件
precaution	*n.*	预防，警惕，防范
carefulness	*n.*	仔细，慎重
amass	*v.*	收集，积聚
reputation	*n.*	名誉，名声
irreparable	*adj.*	不能挽回的
database	*n.*	数据库，资料库
mitigate	*v.*	减轻
countermeasure	*n.*	对策，反措施
eliminate	*v.*	排除，消除
outsourcing	*n.*	外包，外购
evaluate	*vt.*	评价，估计
prudent	*adj.*	谨慎的
definition	*n.*	定义，阐释
attribute	*n.*	属性，品质，特征
portion	*n.*	一部分，一份

organizational	*adj.*	组织的
procedure	*n.*	程序，手续
operator	*n.*	操作员，工作者
confidentiality	*n.*	机密性
disclose	*v.*	揭露，透露
entity	*n.*	实体
integrity	*n.*	完整性
accuracy	*n.*	精确性，正确度
completeness	*n.*	完全
undetected	*adj.*	未被发现的，未被识破的
manner	*n.*	风格，方式，样式
consistency	*n.*	一致性，连贯性
availability	*n.*	可用性，有效性，实用性
non-repudiation	*n.*	不可抵赖性，不可否认性
obligation	*n.*	义务，职责
assist	*v.*	援助，帮助
transcend	*v.*	超越，胜过
realm	*n.*	领域
alter	*v.*	改变
alleged	*adj.*	声称的，所谓的
demonstrate	*v.*	示范，证明，论证
algorithm	*n.*	算法
flaw	*n.*	缺点，瑕疵
	v.	使有缺陷，使无效
invalidate	*v.*	使无效
authenticity	*n.*	确实性，真实性

Phrases

regardless of	不管，不顾
tablet computer	平板电脑
be responsible for…	为……负责
natural disaster	自然灾害
intellectual property	知识产权
phishing attack	钓鱼式攻击
take advantage of	利用
cell phone	手机
confidential information	保密情报，机密信息
financial status	财务状况
in recent years	最近几年中

digital forensics	数字取证
basic principle	基本原理，基本原则
data integrity	数据完整性
referential integrity	参照完整性，引用完整性
transaction processing	事务处理
communication channel	通信电路，信道
shut down	关机，宕机
regard as…	把……认作
digital signature	数字签名
private key	私钥，个人密钥

Abbreviations

InfoSec (Information Security)	安全
CIA (Confidentiality，Integrity and Availability)	保密性、完整性和可用性
ACID (Atomicity, Consistency, Isolation, Durability)	原子性、一致性、隔离性、持久性

Notes

[1] Phishing is the attempt to obtain sensitive information such as usernames, passwords, and credit card details (and, indirectly, money), often for malicious reasons, by disguising as a trustworthy entity in an electronic communication. The word is a neologism created as a homophone of fishing due to the similarity of using a bait in an attempt to catch a victim.

[2] Ransomware is computer malware that installs covertly on a victim's computer, executes a cryptovirology attack that adversely affects it, and demands a ransom payment to decrypt it or not publish it. Simple ransomware may lock the system in a way which is not difficult for a knowledgeable person to reverse, and display a message requesting payment to unlock it. More advanced malware encrypts the victim's files, making them inaccessible, and demands a ransom payment to decrypt them. The ransomware may also encrypt the computer's Master File Table (MFT) or the entire hard drive. Thus, ransomware is a denial-of-access attack that prevents computer users from accessing files since it is intractable to decrypt the files without the decryption key. Ransomware attacks are typically carried out using a Trojan that has a payload disguised as a legitimate file.

[3] A black-hat hacker is a hacker who "violates computer security for little reason beyond maliciousness or for personal gain".

The term was coined by Richard Stallman, to contrast the maliciousness of a criminal hacker versus the spirit of playfulness and exploration of hacker culture, or the ethos of the white-hat hacker, who performs hackerly duties to identify places to repair.

[4] The Gordon-Loeb Model is a mathematical economic model analyzing the optimal investment level in information security. It has been widely referenced in the academic and practitioner literature.

[5] Digital forensics (sometimes known as digital forensic science) is a branch of forensic science encompassing the recovery and investigation of material found in digital devices, often in relation to computer crime.

[6] Referential integrity is a property of data which, when satisfied, requires every value of one attribute (column) of a relation (table) to exist as a value of another attribute (column) in a different (or the same) relation (table).

For referential integrity to hold in a relational database, any column in a base table that is declared a foreign key can contain either a null value, or only values from a parent table's primary key or a candidate key. In other words, when a foreign key value is used it must reference a valid, existing primary key in the parent table. For instance, deleting a record that contains a value referred to by a foreign key in another table would break referential integrity. Some Relational Database Management Systems (RDBMS) can enforce referential integrity, normally either by deleting the foreign key rows as well to maintain integrity, or by returning an error and not performing the delete. Which method is used may be determined by a referential integrity constraint defined in a data dictionary.

The adjective "referential" describes the action that a foreign key performs, "referring" to a link column in another table. In simple terms, "referential integrity" is a guarantee that the target it "refers" to will be found. A lack of referential integrity in a database can lead relational databases to return incomplete data, usually with no indication of an error.

[7] In computer science, transaction processing is information processing that is divided into individual, indivisible operations called transactions. Each transaction must succeed or fail as a complete unit; it can never be only partially complete.

[8] High availability is a characteristic of a system, which aims to ensure an agreed level of operational performance, usually uptime, for a higher than normal period.

Exercises

[Ex. 1] Answer the following questions according to the text.

1. What is information security?
2. What are IT security specialists responsible for?
3. What is a common method of providing information assurance?
4. What are some of the most common information security threats today?
5. What is identity theft?
6. What could happen if confidential information about a business' customers or finances or new product line fell into the hands of a competitor or a black hat hacker?
7. What are the possible responses to a security threat or risk mentioned in the text?
8. What does data integrity mean in information security?
9. What does availability mean?
10. What does non-repudiation imply in law?

[Ex. 2] Translate the following terms or phrases from English into Chinese and vice versa.

1.	transaction processing	1.	
2.	private key	2.	
3.	phishing attack	3.	
4.	digital signature	4.	
5.	data integrity	5.	
6.	算法	6.	
7.	备份；做备份	7.	
8.	机密性	8.	
9.	数据库，资料库	9.	
10.	使无效	10.	

[Ex. 3] Translate the following passage into Chinese.

What Does Information Systems Security Mean?

Information systems security, more commonly referred to as INFOSEC, refers to the processes and methodologies involved in keeping information confidential, available, and assuring its integrity.

It also refers to:

- Access controls, which prevent unauthorized personnel from entering or accessing a system.
- Protecting information no matter where that information is, i.e. in transit (such as in an e-mail) or in a storage area.
- The detection and remediation of security breaches, as well as documenting those events.

Information systems security does not just deal with computer information, but also data and information in all of its forms, such as telephone conversations.

Risk assessments must be performed to determine what information poses the biggest risk. For example, one system may have the most important information on it and therefore will need more security measures to maintain security. Business continuity planning and disaster recovery planning are other facets of an information systems security professional. This professional will plan for what could happen if a major business disruption occurs, but still allow business to continue as usual.

The term is often used in the context of the U.S. Navy, who defines INFOSEC as:

COMPUSEC + COMSEC + TEMPEST = INFOSEC

Where COMPUSEC is computer systems security, COMSEC is communications security, and TEMPEST is compromising emanations.

[Ex. 4] Fill in the blanks with the words given below.

advertising	traffic	mobile	introduced	devices
developing	track	report	alternative	determine

Why Cookies Are Getting Stale

The cookie's days may be numbered. Cookies, the ubiquitous, sometimes innocuous code placed on our computers to help us auto-login to websites, allowing advertisers to ___1___ our online habits, have been a browser mainstay since Lou Montulli of Netscape invented them in 1994.

Because cookies don't work on apps, and third-party cookies are blocked by default in Apple's mobile version of Safari, advertisers have fervently been searching for another way to track ___2___ users. Several companies think they have found a way around cookies' limitations.

Since cookies are not compatible with apps, large tech companies, startups and advertisers are searching for an ___3___ that will allow them to perform cookie-like functions on smartphones and tablets.

The mobile traffic explosion is the reason why tech companies and advertisers feel like cookies should be replaced. In India, mobile generates 61 percent of Internet traffic, according to a recent ___4___ from Statcounter, a firm that helps companies track online visitors. The United States, at 12 percent, is below the worldwide average of 20 percent, according to Statcounter. Still, that's a lot of ___5___ that can't be tracked with cookies by Google and Facebook.

Another holy grail of the post-cookie tech world is the ability to track users across multiple platforms. This is known as cross device ___6___, which is rapidly becoming indispensable for advertisers who want to reach consumers on phones, tablets, desktops and smart TVs.

However, in reality, it's not that easy to track users across myriad ___7___, yet some companies beg to differ. Drawbridge, a San Mateo-based startup, says that they have "matched people to more than 750 million devices," allowing advertisers that use their service to show you, the user, an ad for a trip to Las Vegas on your tablet after "noticing" that during lunch, you searched flights to Vegas from your smartphone.

The company uses proprietary software to match users and their browsing habits to IP addresses to ___8___, with some degree of certainty, that a particular user uses a specific smartphone, tablet and computer.

Experts say that cookies will be around for a while but that the technology has outlived its usefulness, as the little snippets of code are too easily deleted and defeated.

According to a recent article in *USA Today*, Google is ___9___ an anonymous identifier for advertising, or AdID—a "super cookie" of sorts—that would allow advertisers to track online user activities. Apple ___10___ its version of AdID last year for iOS, its mobile operating system.

The *USA Today* article states that a technology like AdID would give users more control, potentially providing the ability to limit tracking via browser settings.

Text B

Risk Management

The Certified Information Systems Auditor[1] (CISA) Review Manual provides the following definition of risk management: "Risk management is the process of identifying vulnerabilities and threats to the information resources used by an organization in achieving business objectives, and

deciding what countermeasures, if any, to take in reducing risk to an acceptable level, based on the value of the information resource to the organization."

There are two things in this definition that may need some clarification. First, the process of risk management is an ongoing, iterative process. It must be repeated indefinitely. The business environment is constantly changing and new threats and vulnerabilities emerge every day. Second, the choice of countermeasures (controls) used to manage risks must strike a balance between productivity, cost, effectiveness of the countermeasure, and the value of the informational asset being protected.

Risk is the likelihood that something bad will happen that causes harm to an informational asset (or the loss of the asset). A vulnerability is a weakness that could be used to endanger or cause harm to an informational asset. A threat is anything (man-made or act of nature) that has the potential to cause harm.

The likelihood that a threat will use a vulnerability to cause harm creates a risk. When a threat does use a vulnerability to inflict harm, it has an impact. In the context of information security, the impact is a loss of availability, integrity, and confidentiality, and possibly other losses (lost income, loss of life, loss of real property). It should be pointed out that it is not possible to identify all risks, nor is it possible to eliminate all risk. The remaining risk is called "residual risk".

1. Risk Assessment

A risk assessment is carried out by a team of people who have knowledge of specific areas of the business. Membership of the team may vary over time as different parts of the business are assessed. The assessment may use a subjective qualitative analysis based on informed opinion, or where reliable dollar figures and historical information is available, the analysis may use quantitative analysis.

The research has shown that the most vulnerable point in most information systems is the human user, operator, designer, or other human. The ISO/IEC 27002:2005[2] Code of practice for information security management recommends the following be examined during a risk assessment:

- security policy;
- organization of information security;
- asset management;
- human resources security;
- physical and environmental security;
- communications and operations management;
- access control;
- information systems acquisition, development and maintenance;
- information security incident management[3] ;
- business continuity management; and
- regulatory compliance.

In broad terms, the risk management process consists of:

- Identification of assets and estimating their value. Include: people, buildings, hardware, software, data (electronic, print, other), and supplies.

- Conduct a threat assessment[4]. Include: Acts of nature, acts of war, accidents, malicious acts originating from inside or outside the organization.
- Conduct a vulnerability assessment, and for each vulnerability, calculate the probability that it will be exploited. Evaluate policies, procedures, standards, training, physical security, quality control, and technical security.
- Calculate the impact that each threat would have on each asset. Use qualitative analysis or quantitative analysis.
- Identify, select and implement appropriate controls. Provide a proportional response. Consider productivity, cost effectiveness, and value of the asset.
- Evaluate the effectiveness of the control measures. Ensure the controls provide the required cost effective protection without discernible loss of productivity.

For any given risk, management can choose to accept the risk based upon the relative low value of the asset, the relative low frequency of occurrence, and the relative low impact on the business. Or, leadership may choose to mitigate the risk by selecting and implementing appropriate control measures to reduce the risk. In some cases, the risk can be transferred to another business by buying insurance or outsourcing to another business. The reality of some risks may be disputed. In such cases leadership may choose to deny the risk.

2. Risk Controls

(1) Administrative Controls

Administrative controls (also called procedural controls) consist of approved written policies, procedures, standards and guidelines. Administrative controls form the framework for running the business and managing people. They inform people on how the business is to be run and how day-to-day operations are to be conducted. Laws and regulations created by government bodies are also a type of administrative control because they inform the business. Some industry sectors have policies, procedures, standards and guidelines that must be followed—the Payment Card Industry Data Security Standard (PCI DSS) required by Visa and MasterCard is such an example. Other examples of administrative controls include the corporate security policy, password policy, hiring policies, and disciplinary policies.

Administrative controls form the basis for the selection and implementation of logical and physical controls. Logical and physical controls are manifestations of administrative controls. Administrative controls are of paramount importance.

(2) Logical Controls

Logical controls (also called technical controls) use software and data to monitor and control access to information and computing systems. For example: Passwords, network and host-based firewalls, network intrusion detection systems, access control lists, and data encryption are logical controls.

An important logical control that is frequently overlooked is the principle of least privilege[5]. The principle of least privilege requires that an individual, program or system process is not granted any more access privileges than are necessary to perform the task. A blatant example of the failure to adhere to the principle of least privilege is logging into Windows as user Administrator to read email and surf the web. Violations of this principle can also occur when an individual collects additional access privileges over time. This happens when employees' job duties change, or they are promoted to a new position, or they transfer to another department. The access privileges required by their new duties are frequently added onto their already existing access privileges which may no longer be necessary or appropriate.

(3) Physical Controls

Physical controls monitor and control the environment of the work place and computing facilities. They also monitor and control access to and from such facilities. For example: Doors, locks, heating and air conditioning, smoke and fire alarms, fire suppression systems, cameras, barricades, fencing, security guards, cable locks, etc. Separating the network and workplace into functional areas are also physical controls.

An important physical control that is frequently overlooked is the separation of duties. Separation of duties ensures that an individual cannot complete a critical task by himself. For example: An employee who submits a request for reimbursement should not also be able to authorize payment or print the check. An applications programmer should not also be the server administrator or the database administrator —these roles and responsibilities must be separated from one another.

3. Defense in Depth

Information security must protect information throughout the life span of the information, from the initial creation of the information on through to the final disposal of the information. The information must be protected while in motion and while at rest. During its lifetime, information may pass through many different information processing systems and through many different parts of information processing systems. There are many different ways the information and information systems can be threatened. To fully protect the information during its lifetime, each component of the information processing system must have its own protection mechanisms. The building up, layering on and overlapping of security measures is called defense in depth. In contrast to a metal chain, which is famously only as strong as its weakest link, the defense in depth aims at a structure where, should one defensive measure fail, other measures will continue to provide protection.

Recall the earlier discussion about administrative controls, logical controls, and physical controls. The three types of controls can be used to form the basis upon which to build a defense in depth strategy. With this approach, defense in depth can be conceptualized as three distinct layers or planes laid one on top of the other. Additional insight into defense in depth can be gained by thinking of it as forming the layers of an onion, with data at the core of the onion, people the next outer layer of the onion, and network security, host-based security and application security forming the outermost layers of the onion. Both perspectives are equally valid and each provides valuable insight into the implementation of a good defense in depth strategy.

4. Security Classification for Information

An important aspect of information security and risk management is recognizing the value of information and defining appropriate procedures and protection requirements for the information. Not all information is equal and so not all information requires the same degree of protection. This requires information to be assigned a security classification.

The first step in information classification is to identify a member of senior management as the owner of the particular information to be classified. Next, develop a classification policy. The policy should describe the different classification labels, define the criteria for information to be assigned a particular label, and list the required security controls for each classification.

Some factors that influence which classification information should be assigned include how much value that information has to the organization, how old the information is and whether or not the information has become obsolete. Laws and other regulatory requirements are also important considerations when classifying information.

The Business Model for Information Security enables security professionals to examine security from systems perspective, creating an environment where security can be managed holistically, allowing actual risks to be addressed.

The type of information security classification labels selected and used will depend on the nature of the organization, with examples being:

- In the business sector, labels such as: Public, Sensitive, Private, and Confidential.
- In the government sector, labels such as: Unclassified, Unofficial, Protected, Confidential, Secret, Top Secret and their non-English equivalents.
- In cross-sectoral formations, the Traffic Light Protocol, which consists of: White, Green, Amber, and Red.

All employees in the organization, as well as business partners, must be trained on the classification schema and understand the required security controls and handling procedures for each classification. The classification of a particular information asset that has been assigned should be reviewed periodically to ensure the classification is still appropriate for the information and to ensure the security controls required by the classification are in place and are followed in their right procedures.

5. Access Control

Access to protected information must be restricted to people who are authorized to access the information. The computer programs, and in many cases the computers that process the information, must also be authorized. This requires that mechanisms be in place to control the access to protected information. The sophistication of the access control mechanisms should be in parity with the value of the information being protected—the more sensitive or valuable the information the stronger the control mechanisms need to be. The foundation on which access control mechanisms are built start with identification and authentication.

Access control is generally considered in three steps: Identification, Authentication, and Authorization.

(1) Identification

Identification is an assertion of who someone is or what something is. If a person makes the statement

"Hello, my name is John Doe" they are making a claim of who they are. However, their claim may or may not be true. Before John Doe can be granted access to protected information it will be necessary to verify that the person claiming to be John Doe really is John Doe. Typically the claim is in the form of a username. By entering that username you are claiming "I am the person the username belongs to".

(2) Authentication

Authentication is the act of verifying a claim of identity. When John Doe goes into a bank to make a withdrawal, he tells the bank teller he is John Doe—a claim of identity. The bank teller asks to see a photo ID, so he hands the teller his driver's license. The bank teller checks the license to make sure it has John Doe printed on it and compares the photograph on the license against the person claiming to be John Doe. If the photo and name match the person, then the teller has authenticated that John Doe is who he claimed to be. Similarly by entering the correct password, the user is providing evidence that he/she is the person the username belongs to.

There are three different types of information that can be used for authentication:

- Something you know: Things such as a PIN, a password, or your mother's maiden name.
- Something you have: A driver's license or a magnetic swipe card.
- Something you are: Biometrics, including palm prints, fingerprints, voice prints and retina (eye) scans.

Strong authentication requires providing more than one type of authentication information (two-factor authentication). The username is the most common form of identification on computer systems today and the password is the most common form of authentication. Usernames and passwords have served their purpose but in our modern world they are no longer adequate. Usernames and passwords are slowly being replaced with more sophisticated authentication mechanisms.

(3) Authorization

After a person, program or computer has successfully been identified and authenticated then it must be determined what informational resources they are permitted to access and what actions they will be allowed to perform (run, view, create, delete, or change). This is called authorization. Authorization to access information and other computing services begins with administrative policies and procedures. The policies prescribe what information and computing services can be accessed, by whom, and under what conditions. The access control mechanisms are then configured to enforce these policies. Different computing systems are equipped with different kinds of access control mechanisms—some may even offer a choice of different access control mechanisms. The access control mechanism a system offers will be based upon one of three approaches to access control or it may be derived from a combination of the three approaches.

The nondiscretionary approach consolidates all access control under a centralized administration. The access to information and other resources is usually based on the individuals function (role) in the organization or the tasks the individual must perform. The discretionary approach gives the creator or owner of the information resource the ability to control access to those resources. In the Mandatory access control approach, access is granted or denied basing upon the security classification assigned to the information resource.

6. Cryptography

Information security uses cryptography to transform usable information into a form that renders it unusable by anyone other than an authorized user; this process is called encryption. Information that has been encrypted (rendered unusable) can be transformed back into its original usable form by an authorized user, who possesses the cryptographic key, through the process of decryption. Cryptography is used in information security to protect information from unauthorized or accidental disclosure while the information is in transit (either electronically or physically) and while information is in storage.

Cryptography provides information security with other useful applications as well including improved authentication methods, message digests, digital signatures, non-repudiation, and encrypted network communications. Older less secure applications such as telnet and ftp are slowly being replaced with more secure applications such as SSH[6] that use encrypted network communications. Wireless communications can be encrypted using protocols such as WPA/WPA2 or the older (and less secure) WEP. Wired communications are secured using AES for encryption and X.1035 for authentication and key exchange. Software applications such as GnuPG[7] or PGP[8] can be used to encrypt data files and e-mail.

Cryptography can introduce security problems when it is not implemented correctly. Cryptographic solutions need to be implemented using industry accepted solutions that have undergone rigorous peer review by independent experts in cryptography. The length and strength of the encryption key is also an important consideration. A key that is weak or too short will produce weak encryption. The keys used for encryption and decryption must be protected with the same degree of rigor as any other confidential information. They must be protected from unauthorized disclosure and destruction and they must be available when needed. Public Key Infrastructure (PKI) solutions address many of the problems that surround key management.

New Words

achieve	*v.*	完成，达到
clarification	*n.*	澄清，净化
iterative	*adj.*	重复的，反复的，迭代的
repeated	*adj.*	重复的，再三的
indefinitely	*adv.*	不确定地
endanger	*v.*	危及
inflict	*v.*	造成
opinion	*n.*	意见，看法，主张，判断，评价
reliable	*adj.*	可靠的，可信赖的
acquisition	*n.*	获得，采集
proportional	*adj.*	比例的，成比例的，相称的，均衡的
discernible	*adj.*	可辨别的
administrative	*adj.*	管理的，行政的

framework	*n.*	构架，框架，结构
manifestation	*n.*	显示，表现
barricade	*v.*	设路障
	n.	路障
disposal	*n.*	处理，处置；布置，安排；配置
conceptualize	*v.*	使有概念，概念化
classification	*n.*	分类，分级
recognize	*v.*	识别，认可，承认
unclassified	*adj.*	无类别的，不保密的
unofficial	*adj.*	非官方的，非法定的，非正式的
schema	*n.*	计划，概要，图表
periodically	*adv.*	周期性地，定时性地
restricted	*adj.*	受限制的
foundation	*n.*	基础，根本
assertion	*n.*	主张，断言，声明
statement	*n.*	声明，陈述
withdrawal	*n.*	收回，退回，取消
evidence	*n.*	证据
retina	*n.*	视网膜
adequate	*adj.*	适当的，足够的
prescribe	*v.*	指示，规定
condition	*n.*	条件，情形
	v.	以……为条件，使达到要求的情况
nondiscretionary	*adj.*	不可任意支配的
consolidate	*v.*	巩固
discretionary	*adj.*	任意的，自由决定的
mandatory	*adj.*	命令的，强制的
possess	*v.*	占有，拥有，持有，支配
rigor	*n.*	严格，严密，精确
surround	*v.*	围绕，包围，环境

Phrases

risk management	风险管理
real property	不动产
point out	指出
residual risk	残留风险
qualitative analysis	定性分析
threat assessment	威胁评估，威胁估计，风险评估
malicious act	恶意行为

quantitative analysis	定量分析
low frequency	低频率
principle of least privilege	最小特权原则(POLP)
fire suppression system	灭火系统
life span	生存期
be trained on…	在……方面接受训练
driver's license	驾驶执照
magnetic swipe card	磁条卡
palm print	掌纹
be derived from	源自
be transformed into	转变为
message digest	信息摘要

Abbreviations

CISA (Certified Information Systems Auditor)	注册信息系统审计师
PIN (Personal Identification Number)	个人身份号码
AES (Advanced Encryption Standard)	高级加密标准

Notes

[1] Certified Information Systems Auditor (CISA) is a globally recognized certification in the field of audit, control and security of information systems. CISA gained worldwide acceptance having uniform certification criteria, the certification has a high degree of visibility and recognition in the fields of IT security, IT audit, IT risk management and governance. Vacancies in the areas of IT security management, IT audit or IT risk management often ask for a CISA certification.

[2] ISO/IEC 27002 is an information security standard published by the International Organization for Standardization (ISO) and by the International Electrotechnical Commission (IEC).

[3] Incident Management (IcM) is a term describing the activities of an organization to identify, analyze, and correct hazards to prevent a future re-occurrence. These incidents within a structured organization are normally dealt with by either an Incident Response Team (IRT), or an Incident Management Team (IMT). These are often designated before hand, or during the event and are placed in control of the organization whilst the incident is dealt with, to restore normal functions.

[4] Threat assessment comprises strategies or pathways used to determine the credibility and seriousness of a potential threat, as well as the likelihood that it will be carried out in the future. Threat assessment is a violence prevention act that involves: identifying the threats to commit violent act; determining seriousness of threat; developing intervention plans that protect potential victims and address underlying problem that stimulated the threatening behaviour.

[5] The principle of least privilege (POLP; also known as the principle of least authority) is an important concept in computer security, promoting minimal user profile privileges on computers, based on users' job necessities. It can also be applied to processes on the computer; each system component or

process should have the least authority necessary to perform its duties. This helps reduce the "attack surface" of the computer by eliminating unnecessary privileges that can result in network exploits and computer compromises. It is widely recognized as an important design consideration in enhancing the protection of data and functionality from faults (fault tolerance) and malicious behavior.

[6] Secure Shell (SSH) is a cryptographic network protocol for operating network services securely over an unsecured network. The best known example application is for remote login to computer systems by users.

SSH provides a secure channel over an unsecured network in a client-server architecture, connecting an SSH client application with an SSH server. Common applications include remote command-line login and remote command execution, but any network service can be secured with SSH.

[7] GNU Privacy Guard (GnuPG or GPG) is a free software replacement for Symantec's PGP cryptographic software suite.

[8] Pretty Good Privacy (PGP) is an encryption program that provides cryptographic privacy and authentication for data communication. PGP is often used for signing, encrypting, and decrypting texts, e-mails, files, directories, and whole disk partitions and to increase the security of e-mail communications.

Exercises

[Ex. 5] Answer the following questions according to the text.

1. What is the definition of risk management provided by the Certified Information Systems Auditor (CISA) Review Manual?

2. What are risk, a vulnerability and a threat respectively?

3. What kind of analysis may the risk assessment use?

4. What do administrative controls consist of? What do they form? And what do they inform people on?

5. What do logical controls use to monitor and control access to information and computing systems? What is an important logical control that is frequently overlooked?

6. What do physical controls monitor and control? What is an important physical control that is frequently overlooked?

7. What is called defense in depth? What does it aim at?

8. What are the steps in information classification?

9. How many steps is access control generally considered? What are they respectively?

10. What is cryptography used to do in information security?

Reading Material

Information Security Management System

An information security management system (ISMS) is a set of policies concerned with information security management or IT related risks.

The governing principle[1] behind an ISMS is that an organization should design, implement and maintain a coherent[2] set of policies, processes and systems to manage risks to its information assets, thus ensuring acceptable[3] levels of information security risk.

1. ISMS Description

As with all management processes, an ISMS must remain effective and efficient in the long term, adapting to changes in the internal organization and external environment. ISO/IEC 27001:2005 therefore incorporated the "Plan-Do-Check-Act" (PDCA[4]) (see Figure 3-1), or Deming cycle, approach:

- The "Plan" phase is about designing the ISMS, assessing information security risks and selecting appropriate controls.
- The "Do" phase involves implementing and operating the controls.
- The objective of the "Check" phase is to review and evaluate the performance (efficiency and effectiveness[5]) of the ISMS.
- In the "Act" phase, changes are made where necessary to bring the ISMS back to peak performance.

ISO/IEC 27001:2005 is a risk-based information security standard, which means that organizations need to have a risk management process in place. The risk management process fits into[6] the PDCA model given above.

However, the latest standard, ISO/IEC 27001:2013, does not emphasize[7] the Deming cycle anymore. The ISMS user is free to use any management process (improvement) approach like PDCA or Six Sigmas DMAIC[8].

Figure 3-1 Plan-Do-Check-Act Cycle

---

1 governing principle 指导原则

2 coherent *adj.* 一致的，连贯的

3 acceptable *adj.* 可接受的

4 PDCA 循环又称质量环，是管理学中的一个通用模型，最早由休・哈特于 1930 年构想，后来被美国质量管理专家戴明博士在 1950 年再度挖掘出来，并加以广泛宣传，运用于持续改善产品质量的过程。

5 effectiveness *n.* 效力

6 fit into 适合

7 emphasize *v.* 强调，着重

8 DMAIC 是六西格玛管理中流程改善的重要工具。DMAIC 是指 Define(定义)、Measure (测量)、Analyze (分析)、Improve (改进)、Control(控制)五个阶段。

2. Need for an ISMS

Security experts say:

- Information technology security administrators should expect to devote approximately one-third of their time addressing technical aspects[1]. The remaining two-thirds should be spent developing policies and procedures, performing security reviews and analyzing risk, addressing contingency[2] planning and promoting security awareness.
- Security depends on people more than on technology.
- Employees are a far greater threat to information security than outsiders[3].
- Security is like a chain. It is only as strong as its weakest link.
- The degree[4] of security depends on three factors: The risk you are willing to take, the functionality of the system and the costs you are prepared to pay.
- Security is not a status or a snapshot, but a running process.

These facts inevitably[5] lead to the conclusion that security administration is a management issue, and not a purely technical issue.

The establishment, maintenance and continuous[6] update of an ISMS provide a strong indication that a company is using a systematic[7] approach for the identification, assessment and management of information security risks.

The following are the critical factors[8] of ISMS:

- Confidentiality: Protecting information from unauthorized parties.
- Integrity: Protecting information from modification by unauthorized users.
- Availability: Making the information available to authorized users.

A company will be capable of[9] successfully addressing information Confidentiality, Integrity and Availability (CIA) requirements which in turn have implications:

- business continuity;
- minimization of damages and losses;
- competitive edge[10];
- profitability and cash-flow;
- respected organization image;
- legal compliance.

The chief objective of information security management is to implement the appropriate

---

1 technical aspect 技术现状
2 contingency *n.* 偶然，可能性
3 outsider *n.* 局外人
4 degree *n.* 度，程度
5 inevitably *adv.* 不可避免
6 continuous *adj.* 连续的，持续的
7 systematic *adj.* 系统的，体系的
8 critical factor 关键因素
9 be capable of 能够
10 competitive edge 竞争优势

measurements in order to eliminate or minimize the impact that various security related threats and vulnerabilities might have on an organization. In doing so, information security management will enable implementing the desirable qualitative characteristics[1] of the services offered by the organization (i.e. availability of services, preservation[2] of data confidentiality and integrity etc.). By preventing and minimizing the impacts of security incidents[3], ISMS ensures business continuity, customer confidence, protect business investments and opportunities, or reduce damage to the business.

Large organizations, banks and financial institutes, telecommunication operators[4], hospital and health institutes and public or governmental bodies have many reasons to address information security very seriously. Legal and regulatory requirements which aim at protecting sensitive or personal data as well as general public security requirements impel them to devote the utmost[5] attention and priority to information security risks.

Under these circumstances, the development and implementation of a separate and independent management process—namely an ISMS—is the only alternative.

The development of an ISMS framework based on ISO/IEC 27001:2005 entails[6] the following following six steps:

- definition of security policy;
- definition of ISMS scope;
- risk assessment (as part of risk management) ;
- risk management;
- selection of appropriate controls;
- statement of applicability[7].

3. Critical Success Factors for ISMS

To be effective, the ISMS must:

- Have the continuous, unshakable[8] and visible support and commitment of the organization's organization's top management.
- Be managed centrally, based on a common strategy and policy across the entire organization.
- Be an integral part of the overall management of the organization related to and reflecting the organization's approach to risk management, the control objectives and controls and the degree of assurance required.
- Have security objectives and activities based on business objectives and requirements and led by business management.

1 qualitative characteristics　质量特征
2 preservation　*n.*　保存
3 incident　*n.*　事件
4 telecommunication operator　电信营运商
5 to the utmost　极度
6 entail　*v.*　使必需，使承担
7 applicability　*n.*　适用性，适应性
8 unshakable　*adj.*　不可动摇的，坚定不移的

- Undertake[1] only necessary tasks and avoiding over-control and waste of valuable resources.
- Fully comply with the organization philosophy and mindset[2] by providing a system that instead of preventing people from doing what they are employed to do, it will enable them to do it in control and demonstrate their fulfilled accountabilities[3].
- Be based on continuous training and awareness of staff and avoid the use of disciplinary[4] measures and "police" or "military" practices.
- Be a never ending process.

4. Dynamic Issues in ISMS

There are three main problems which lead to uncertainty[5] in Information Security Management Systems (ISMS):

(1) Dynamically Changing Security Requirements of an Organization

Rapid technological development raises new security concerns for organizations. The existing security measures and requirements become obsolete[6] as new vulnerabilities arise with the development in technology. To overcome this issue, the ISMS should organize and manage dynamically changing requirements and keep the system up-to-date.

(2) Externalities Caused by a Security System

Externality[7] is an economic concept for the effects borne by the party that is not directly involved in a transaction. Externalities could be positive or negative. The ISMS deployed in an organization may also cause externalities for other interacting systems. Externalities caused by the ISMS are uncertain and cannot be predetermined before the ISMS is deployed. The internalization[8] of externalities caused by the ISMS is needed in order to benefit internalizing[9] organizations and interacting partners by protecting them from vulnerable ISMS behaviors.

(3) Obsolete Evaluation of Security Concerns

The evaluations of security concerns used in ISMS become obsolete as the technology progresses and new threats and vulnerabilities arise. The need for continuous security evaluation of organizational products, services, methods and technology is essential to[10] maintain an effective ISMS. The evaluated security concerns need to be re-evaluated. A continuous security evaluation mechanism of ISMS within the organization is a critical need to achieve information security objectives. The re-evaluation process is tied with dynamic security requirement management process discussed above.

---

1 undertake *v.* 承担，担任

2 mindset *n.* 思维方式，观念

3 accountability *n.* 职责，有责任，有义务

4 disciplinary *adj.* 训练的

5 uncertainty *n.* 不确定，不可靠

6 obsolete *adj.* 荒废的，陈旧的

7 externality *n.* 外部性 外部性又称溢出效应、外部影响、外差效应或外部效应、外部经济，是指一个人或一群人的行动和决策使另一个人或一群人受损或受益的情况。

8 internalization *n.* 内在化

9 internalize *v.* 使内在化

10 be essential to 对……是必要的

参考译文

信息安全

信息安全（有时缩写为 InfoSec）是防止未经授权的访问、使用、披露、中断、修改、检查、记录或破坏信息的行为。无论数据为何种形式（如电子、物理），都可以使用这一术语。

1. 概述

（1）信息技术（IT）安全

IT 安全有时也称为计算机安全，信息技术安全是把信息安全应用于技术（通常是某种形式的计算机系统）。值得注意的是，计算机并不一定意味着是个人计算机。计算机是指具有处理器和一些内存的任何设备。这样的设备的范围可以从未联网的独立设备（如像计算机器这样的简单设备）到网络移动计算设备（如智能电话和平板电脑）。由于大型企业数据的性质和价值，几乎在所有主要企业/机构中都能看到 IT 安全专家。他们负责保护公司内部的所有技术，防范那些试图获取关键私人信息或获取内部系统控制的恶意网络攻击。

提供信息信任的行为，就是保证信息的保密性、完整性和可用性（CIA）不受侵犯。例如，确保出现关键问题时数据不会丢失。这些问题包括但不限于自然灾害、计算机/服务器故障或物理失窃。现如今，由于大多数信息存储在计算机上，信息保障通常由 IT 安全专家负责。提供信息保障的常见方法是在出现上述问题之一时对数据进行异地备份。

（2）威胁

信息安全威胁有许多形式。如今最常见的一些威胁是软件攻击、盗窃知识产权、窃取身份、窃取信息、破坏和信息敲诈。大多数人都经历过某种软件攻击。病毒、蠕虫、网络钓鱼攻击和特洛伊木马是软件攻击的一些常见例子。盗用知识产权也是 IT 领域中许多企业遇到的一个广泛问题。身份盗用是企图以别人的身份行事，以获取该人的个人信息或利用其获取重要信息。由于当今大多数设备是移动设备，因此信息的窃取变得越来越普遍。破坏通常包括破坏组织的网站，其目的是让其客户失去信心。信息勒索包括盗窃公司的财产或信息，如果公司要换回信息或财产，就要支付报酬（如使用勒索软件）。有许多方法可以帮助保护自己免受这些攻击，但最为有用的预防措施之一就是用户一定要细心。

政府、军队、公司、金融机构、医院和私营企业都聚集了大量的关于员工、客户、产品、研究和财务状况的机密信息。现在这些信息大部分由电子计算机收集、处理和存储，并通过网络传输给其他计算机。

如果企业的客户、财务或新产品线的秘密信息落在竞争对手或黑帽黑客手中，企业及其客户可能遭受广泛的、无法弥补的经济损失并有损于公司声誉。从商业角度来看，信息安全必须与成本平衡，Gordon-Loeb 模型为解决这一问题提供了一种数学经济方法。

对于个人而言，信息安全对隐私具有重大影响，在不同的文化中，对信息安全的看法截然不同。

信息安全领域近年来不断发展壮大。它提供了许多专业领域，包括保护网络和盟军基础设施、确保应用程序和数据库安全、安全测试、信息系统审计、业务连续性规划和数字取证。

对安全威胁或风险的可能回应是：

- 减少/减轻——实施安全措施和对策，以消除漏洞或阻止威胁。
- 分配/转移——将受威胁的成本放到另一个实体或组织上，如购买保险或外包。
- 接受——评估对策成本是否超过威胁导致的损失成本。
- 忽略/拒绝——这不是一个有效的或谨慎妥当的回应。

2. 定义

信息安全的属性或特点，即保密性、完整性和可用性（CIA）。信息系统由符合信息安全行业标准的硬件、软件和通信三个部分组成。其保护和预防机制有：物理、个人和组织三个级别或层次。通常通过实施程序或策略以确保组织内的信息安全。

3. 基本原则

（1）保密

在信息安全方面，保密性的特点是指未经授权的个人或实体不能使用信息，或者不能把信息披露给他们。

（2）完整性

在信息安全方面，数据完整性意味着在整个生命周期内维护和保证数据的准确性和完整性。这意味着无法以未授权或未被发现的方式修改数据。这与数据库中的引用完整性不一样，尽管它可以被视为一种特殊的一致性情况，就像理解事务处理的经典 ACID 模型那样。信息安全系统通常还提供消息完整性以及数据机密性。

（3）可用性

任何信息系统要达到其目的，信息就必须在需要时可用。这意味着用于存储和处理信息的计算机系统、用于保护信息的安全控制以及用于访问信息的通信信道都必须正常工作。高可用性系统的目的是始终保持系统可用，防止由于停电、硬件故障和系统升级引起的服务中断。确保可用性还涉及防止拒绝服务攻击，例如大量涌入消息迫使目标系统关闭。

（4）不可否认性

在法律上，不可否认意味着某人打算履行合同中的义务。这也意味着交易的一方不能否认已经收到的交易，另一方也不能否认已经发送的交易。注意：这也被认为是完整性的一部分。

要注意，重要的是，虽然诸如加密系统之类的技术可以帮助实现不可否认性，但这个概念的核心是超越技术领域的法律概念。例如，这不足以表明消息与发送者的私钥签名的数字签名相匹配，因此只要发送者已经发送了消息，没有人可以在传输过程中更改它。所谓的发件人也可以表明数字签名算法是脆弱的或有缺陷的，或指称或证明他的签名密钥已被泄密。这些违规行为的错误可能与发件人有涉或无涉，这种说法可能会或可能不会减轻发件人的责任，但是这个断言可以使证明真实性和完整性的签名的声明无效，从而防止抵赖。

Unit 4

Text A

Security Hackers

A security hacker is someone who seeks to breach defenses and exploit weaknesses in a computer system or network. Hackers may be motivated by a multitude of reasons, such as profit, information gathering, challenge, recreation, or to evaluate system weaknesses to assist in formulating defenses against potential hackers.

1. Classifications

Several subgroups of the computer underground with different attitudes use different terms to demarcate themselves from each other, or try to exclude some specific group with whom they do not agree.

Eric S. Raymond, author of *The New Hacker's Dictionary*, advocates that members of the computer underground should be called crackers. Yet, those people see themselves as hackers and even try to include the views of Raymond in what they see as a wider hacker culture, a view that Raymond has harshly rejected. Instead of a hacker/cracker dichotomy, they emphasize a spectrum of different categories, such as white hat, grey hat, black hat and script kiddie. In contrast to Raymond, they usually reserve the term cracker for more malicious activity.

According to Ralph D. Clifford, a cracker or cracking is to "gain unauthorized access to a computer in order to commit another crime such as destroying information contained in that system". These subgroups may also be defined by the legal status of their activities.

(1) White Hat

A white hat hacker breaks security for non-malicious reasons, either to test their own security system, perform penetration tests or vulnerability assessments for a client—or while working for a security company which makes security software. The term is generally synonymous with ethical hacker, and the EC-Council, among others, have developed certifications, courseware, classes, and online training covering the diverse arena of ethical hacking.

(2) Black Hat

A black hat hacker is a hacker who violates computer security for little reason beyond maliciousness or for personal gain. The term was coined by Richard Stallman, to contrast the maliciousness of a criminal hacker versus the spirit of playfulness and exploration in hacker culture, or the ethos of the white hat hacker who performs hacking duties to identify places to repair or as a means of legitimate employment. Black hat hackers form the stereotypical, illegal hacking groups often portrayed in popular culture, and are "the epitome of all that the public fears in a computer criminal".

(3) Grey Hat

A grey hat hacker lies between a black hat and a white hat hacker. A grey hat hacker may surf the Internet and hack into a computer system for the sole purpose of notifying the administrator that their system has a security defect, for example. They may then offer to correct the defect for a fee. Grey hat hackers sometimes find the defect of a system and publish the facts to the world instead of a group of people. Even though grey hat hackers may not necessarily perform hacking for their personal gain, unauthorized access to a system can be considered illegal and unethical.

(4) Elite Hacker

A social status among hackers, elite is used to describe the most skilled. Newly discovered exploits circulate among these hackers. Elite groups such as Masters of Deception conferred a kind of credibility on their members.

(5) Script Kiddie

A script kiddie (also known as a skid or skiddie) is an unskilled hacker who breaks into computer systems by using automated tools written by others (usually by other black hat hackers), hence the term script (i.e. a prearranged plan or set of activities) kiddie (i.e. kid, child—an individual lacking knowledge and experience, immature), usually with little understanding of the underlying concept.

(6) Blue Hat

A blue hat hacker is someone outside computer security consulting firms who is used to bug-test a system prior to its launch, looking for exploits so they can be closed. Microsoft also uses the term BlueHat to represent a series of security briefing events.

(7) Hacktivist

A hacktivist is a hacker who utilizes technology to publicize a social, ideological, religious or political message.

Hacktivism can be divided into two main groups:

1) Cyberterrorism: Activities involving website defacement or denial-of-service attacks.

2) Freedom of information: Making information that is not public, or is public in non-machine-readable formats, accessible to the public.

2. Attacks

Typical approaches in an attack on Internet-connected system are:

1) Network enumeration[1]: Discovering information about the intended target.

2) Vulnerability analysis: Identifying potential ways of attack.

3) Exploitation: Attempting to compromise the system by employing the vulnerabilities found through the vulnerability analysis.

In order to do so, there are several recurring tools of the trade and techniques used by computer criminals and security experts.

(1) Security Exploits

A security exploit is a prepared application that takes advantage of a known weakness. Common examples of security exploits are SQL injection[2], Cross-Site Scripting(XSS)[3] and Cross-Site Request Forgery(CSRF)[4] which abuse security holes that may result from substandard programming practice. Other exploits would be able to be used through File Transfer Protocol (FTP), Hypertext Transfer Protocol (HTTP), PHP, SSH[5], Telnet[6] and some Web pages. These are very common in Web site and Web domain hacking.

(2) Techniques

1) Vulnerability scanner. A vulnerability scanner is a tool used to quickly check computers on a network for known weaknesses. Hackers also commonly use port scanners. These check to see which ports on a specified computer are "open" or available to access the computer, and sometimes will detect what program or service is listening on that port, and its version number. (Firewalls defend computers from intruders by limiting access to ports and machines, but they can still be circumvented.)

2) Finding vulnerabilities. Hackers may also attempt to find vulnerabilities manually. A common approach is to search for possible vulnerabilities in the code of the computer system then test them, sometimes reverse engineering[7] the software if the code is not provided.

3) Brute-force attack. Password guessing. This method is very fast when used to check all short passwords, but for longer passwords other methods such as the dictionary attack are used, because of the time a brute-force search takes.

4) Password cracking. Password cracking is the process of recovering passwords from data that has been stored in or transmitted by a computer system. Common approaches include repeatedly trying guesses for the password, trying the most common passwords by hand, and repeatedly trying passwords from a "dictionary", or a text file with many passwords.

5) Packet analyzer. A packet analyzer ("packet sniffer") is an application that captures data packets, which can be used to capture passwords and other data in transit over the network.

6) Spoofing attack (phishing). A spoofing attack involves one program, system or website that successfully masquerades as another by falsifying data and is thereby treated as a trusted system by a user or another program—usually to fool programs, systems or users into revealing confidential information, such as user names and passwords.

7) Rootkit[8]. A rootkit is a program that uses low-level, hard-to-detect methods to subvert control of an operating system from its legitimate operators. Rootkits usually obscure their installation and attempt to prevent their removal through a subversion of standard system security. They may include replacements for system binaries, making it virtually impossible for them to be detected by checking process tables.

8) Trojan horses. A Trojan horse is a program that seems to be doing one thing but is actually doing another. It can be used to set up a backdoor[9] in a computer system, enabling the intruder to

gain access later. (The name refers to the horse from the Trojan War, with the conceptually similar function of deceiving defenders into bringing an intruder into a protected area.)

9) Computer virus. A virus is a self-replicating program that spreads by inserting copies of itself into other executable code or documents. By doing this, it behaves similarly to a biological virus, which spreads by inserting itself into living cells. While some viruses are harmless or mere hoaxes, most are considered malicious.

10) Computer worm. Like a virus, a worm is also a self-replicating program. It differs from a virus in that it propagates through computer networks without user intervention; and it does not need to attach itself to an existing program. Nonetheless, many people use the terms "virus" and "worm" interchangeably to describe any self-propagating program.

11) Keystroke logging. A keylogger is a tool designed to record ("log") every keystroke on an affected machine for later retrieval, usually to allow the user of this tool to gain access to confidential information typed on the affected machine. Some keyloggers use virus-, trojan-, and rootkit-like methods to conceal themselves. However, some of them are used for legitimate purposes, even to enhance computer security. For example, a business may install a keylogger on a computer used at a point of sale to detect evidence of employee fraud.

New Words

hacker	*n.*	黑客
exploit	*v.*	使用
weakness	*n.*	弱点，漏洞
multitude	*n.*	大量
challenge	*n.*	挑战
	v.	向……挑战
recreation	*n.*	消遣，娱乐
formulate	*v.*	构想出，规划，确切地阐述，用公式表示
defense	*n. & v.*	防卫
potential	*adj.*	潜在的，可能的
subgroup	*n.*	子类，子群
demarcate	*v.*	划分界线
exclude	*v.*	把……排除在外
advocate	*n.*	提倡者，鼓吹者
	v.	提倡，鼓吹
harshly	*adv.*	严厉地，苛刻地
dichotomy	*n.*	两分，二分法；分裂
spectrum	*n.*	光谱，型谱
reserve	*v.*	保留，预定
synonymous	*adj.*	同义的
playfulness	*n.*	玩兴，嬉戏，活泼快乐，戏谑，童心

exploration	*n.* 探测，踏勘；研究，探索
ethos	*n.* 文化精神，社会的特质，气质，民族精神，社会思潮，社会或团体的生活准则
stereotypical	*adj.* 典型的，带有成见的，老套的
illegal	*adj.* 违法的，非法的
defect	*n.* 过失，缺点
unethical	*adj.* 不道德的，缺乏职业道德的
elite	*n.* 精英，精锐，中坚分子
credibility	*n.* 可信性
unskilled	*adj.* 不熟练的，没有技巧的，无须技能的
script	*n.* 脚本
prearrange	*v.* 预先安排
immature	*adj.* 不成熟的，未完全发展的
hacktivist	*n.* 黑客主义者
publicize	*v.* 宣传，宣扬，推广，传播
ideological	*adj.* 意识形态的
religious	*adj.* 宗教上的
cyberterrorism	*n.* 网络恐怖主义
defacement	*v.* 丑化，损毁，涂改
attack	*n. & v.* 攻击
approach	*n.* 方法，步骤，途径
enumeration	*n.* 列举
compromise	*v.* 危及……的安全，违背（原则）
	n. 损害，妥协，折中方案
recur	*v.* 复发，重现，再来
expert	*n.* 专家，行家
	adj. 老练的，内行的，专门的
	v. 在……中当行家，当专家
injection	*n.* 注入，注射
circumvent	*v.* 设法避免，设法规避
manually	*adv.* 手动地，人工地
packet	*n.* 数据包
sniffer	*n.* 嗅探器
falsify	*v.* 伪造，篡改
treat	*v.* 对待，招待；处理，加工
rootkit	*n.* 引导工具，根工具包
subvert	*v.* 推翻，暗中破坏，搅乱
obscure	*adj.* 暗的，朦胧的，模糊的
	v. 使暗，使不明显

removal	*n.* 删除，移动
subversion	*n.* 破坏，颠覆
deceive	*v.* 欺骗，行骗
replicate	*v.* 复制
insert	*v.* 插入，嵌入
executable	*adj.* 可执行的
document	*n.* 文件，文档 *v.* 证明，记录
harmless	*adj.* 无害的
hoax	*v.* & *n.* 愚弄
propagate	*v.* 繁殖，传播
intervention	*n.* 干涉
attach	*v.* 缚上，系上，贴上
interchangeably	*adv.* 可交换地
keystroke	*n.* 键击，按键
keylogger	*n.* 键盘记录器
retrieval	*n.* 取回，恢复
conceal	*v.* 隐藏，隐蔽，隐瞒
fraud	*n.* 欺骗，欺诈行为，诡计，骗子

Phrases

information gathering	搜集信息，收集情报
hacker culture	黑客文化
white hat	白帽
grey hat	灰帽
black hat	黑帽
in contrast to…	和……形成对比，和……形成对照
penetration test	渗透测试
vulnerability assessment	漏洞评估
ethical hacker	道德黑客
for the purpose of	为了
automated tool	自动工具
blue hat	蓝帽
bug-test a system	错误测试系统
website defacement	网站涂改
intended target	指定目标
in order to	为了
cross-site scripting(XSS)	跨站脚本
cross-site request forgery(CSRF)	跨站请求伪造

vulnerability scanner	漏洞扫描器
port scanner	端口扫描器，端口扫描程序
version number	版本号
search for	搜索，搜寻
reverse engineering	逆向工程，逆向技术
brute-force search	暴力搜索，蛮力搜索
text file	文本文件
packet sniffer	包监听器，包探测器
Trojan horse	特洛伊木马
set up	设立，建立
back door	后门
self-replicating program	自复制程序
point of sale	销售点

Abbreviations

SQL (Structured Query Language)	结构化查询语言
FTP (File Transfer Protocol)	文件传输协议
HTTP (Hypertext Transfer Protocol)	超文本传输协议
SSH (Secure Shell)	安全外壳协议

Notes

[1] Network enumeration is a computing activity in which usernames and info on groups, shares, and services of networked computers are retrieved. It should not be confused with network mapping, which only retrieves information about which servers are connected to a specific network and what operating system runs on them.

[2] SQL injection is a code injection technique, used to attack data-driven applications, in which nefarious SQL statements are inserted into an entry field for execution (e.g. to dump the database contents to the attacker). SQL injection must exploit a security vulnerability in an application's software, for example, when user input is either incorrectly filtered for string literal escape characters embedded in SQL statements or user input is not strongly typed and unexpectedly executed. SQL injection is mostly known as an attack vector for websites but can be used to attack any type of SQL database.

[3] Cross-Site Scripting (XSS) is a type of computer security vulnerability typically found in web applications. XSS enables attackers to inject client-side scripts into web pages viewed by other users. A cross-site scripting vulnerability may be used by attackers to bypass access controls such as the same-origin policy. Cross-site scripting carried out on websites accounted for roughly 84% of all security vulnerabilities documented by Symantec as of 2007. Bug bounty company HackerOne in 2017 reported that XSS is still a major threat vector. XSS effects vary in range from petty nuisance to significant security risk, depending on the sensitivity of the data handled by the vulnerable site

and the nature of any security mitigation implemented by the site's owner.

[4] Cross-Site Request Forgery, also known as one-click attack or session riding and abbreviated as CSRF (sometimes pronounced sea-surf) or XSRF, is a type of malicious exploit of a website where unauthorized commands are transmitted from a user that the web application trusts. Unlike cross-site scripting (XSS), which exploits the trust a user has for a particular site, CSRF exploits the trust that a site has in a user's browser.

[5] Secure Shell (SSH) is a cryptographic network protocol for operating network services securely over an unsecured network. The best known example application is for remote login to computer systems by users.

SSH provides a secure channel over an unsecured network in a client-server architecture, connecting an SSH client application with an SSH server. Common applications include remote command-line login and remote command execution, but any network service can be secured with SSH. The protocol specification distinguishes between two major versions, referred to as SSH-1 and SSH-2.

[6] Telnet is a protocol used on the Internet or local area networks to provide a bidirectional interactive text-oriented communication facility using a virtual terminal connection. User data is interspersed in-band with Telnet control information in an 8-bit byte oriented data connection over the Transmission Control Protocol (TCP).

[7] Reverse engineering, also called back engineering, is the processes of extracting knowledge or design information from anything man-made and reproducing it or reproducing anything based on the extracted information. The process often involves disassembling something (a mechanical device, electronic component, computer program, or biological, chemical, or organic matter) and analyzing its components and workings in detail.

[8] A rootkit is a collection of computer software, typically malicious, designed to enable access to a computer or areas of its software that would not otherwise be allowed (for example, to an unauthorized user) and often masks its existence or the existence of other software. The term rootkit is a concatenation of "root" (the traditional name of the privileged account on Unix-like operating systems) and the word "kit" (which refers to the software components that implement the tool). The term "rootkit" has negative connotations through its association with malware.

[9] A backdoor is a method, often secret, of bypassing normal authentication or encryption in a computer system, a product, or an embedded device (e.g. a home router), or its embodiment, e.g. as part of a cryptosystem, an algorithm, a chipset, or a "homunculus computer" (such as that as found in Intel's AMT technology). Backdoors are often used for securing remote access to a computer, or obtaining access to plaintext in cryptographic systems.

A backdoor may take the form of a hidden part of a program one uses, a separate program (e.g. Back Orifice may subvert the system through a rootkit), or code in the firmware of ones hardware or parts of ones operating system such as Microsoft Windows. Although normally surreptitiously installed, in some cases backdoors are deliberate and widely known. These kinds of backdoors might have “legitimate” uses such as providing the manufacturer with a way to restore user passwords.

Exercises

[Ex. 1] Answer the following questions according to the text.

1. What is a security hacker?
2. What does a white hat hacker do?
3. What is a black hat hacker? Who coined the term?
4. What may a grey hat hacker do?
5. What is a script kiddie?
6. How many main groups can hacktivists be divided into? What are they?
7. What are the typical approaches in an attack on Internet-connected system mentioned in the text?
8. What is password cracking? What does the common approaches include?
9. What does a spoofing attack involve?
10. What is a keylogger?

[Ex. 2] Translate the following terms or phrases from English into Chinese and vice versa.

1. vulnerability scanner	1. ________
2. port scanner	2. ________
3. website defacement	3. ________
4. packet sniffer	4. ________
5. information gathering	5. ________
6. 文件，文档；证明，记录	6. ________
7. 可执行的	7. ________
8. 探测，踏勘；研究，探索	8. ________
9. 可交换地	9. ________
10. 数据包	10. ________

[Ex. 3] Translate the following passage into Chinese.

DoS

In computing, a Denial-of-Service attack (DoS attack) is a cyber-attack where the perpetrator seeks to make a machine or network resource unavailable to its intended users by temporarily or indefinitely disrupting services of a host connected to the Internet. Denial of service is typically accomplished by flooding the targeted machine or resource with superfluous requests in an attempt to overload systems and prevent some or all legitimate requests from being fulfilled. A DoS attack is analogous to a group of people crowding the entry door or gate to a shop or business, and not letting legitimate parties enter into the shop or business, disrupting normal operations.

In a DoS attack, the incoming traffic flooding the victim originates from many different sources —potentially hundreds of thousands or more. This effectively makes it impossible to stop the attack simply by blocking a single IP address; plus, it is very difficult to distinguish legitimate user traffic from attack traffic when spread across so many points of origin. There are two general forms of DoS attacks: Those that crash services and those that flood services. The most serious attacks are distributed. Many attacks involve forging of IP sender addresses (IP address spoofing) so that the location of the attacking machines cannot easily be identified and so that the attack cannot be easily defeated using ingress filtering.

[Ex. 4] Fill in the blanks with the words given below.

attack	modification	Categories	demonstrated	harm
destroy	confidential	protection	vulnerabilities	target

Cyber Attacks

A cyber attack is a digital invasion, usually caused by hackers, made through the Internet or a computer network. This may be done to access __1__ information or to destroy the network itself with the use of malware. Cyber attacks may not always be easily detected, especially if the goal is simply to access information.

Cyber attacks can range from installing spyware on a PC to attempts to __2__ the infrastructure of entire nations. Cyber attacks have become increasingly sophisticated and dangerous as the Stuxnet worm recently __3__. User behavior analytics and SIEM are used to prevent these attacks. Legal experts are seeking to limit use of the term to incidents causing physical damage, distinguishing it from the more routine data breaches and broader hacking activities.

Professional hackers, either working on their own or employed by the government or military service, can find computer systems with __4__ lacking the appropriate security software. Once found, they can infect systems with malicious code and then remotely control the system or computer by sending commands to view content or to disrupt other computers. There needs to be a pre-existing system flaw within the computer such as no antivirus __5__ or faulty system configuration for the viral code to work. Many professional hackers will promote themselves to cyber terrorists where a new set of rules govern their actions. Cyber terrorists have premeditated plans and their attacks are not born of rage. They need to develop their plans step-by-step and acquire the appropriate software to carry out an __6__. They usually have political agendas, targeting political structures. Cyber terrorists are hackers with a political motivation, their attacks can impact political structure through this corruption and destruction. They also __7__civilians, civilian interests and civilian installations. As previously stated cyber terrorists attack persons or property and cause enough __8__ to generate fear.

In detail, there are a number of techniques to utilize in cyber attacks and a variety of ways to administer them to individuals or establishments on a broader scale. Attacks are broken down into two __9__: syntactic attacks and semantic attacks. Syntactic attacks are straightforward; it is considered malicious software which includes viruses, worms, and Trojan horses. Semantic attack is the __10__ and dissemination of correct and incorrect information. Information modified could have been done without the use of computers even though new opportunities can be found by using them. To set someone into the wrong direction or to cover your tracks, the dissemination of incorrect information can be utilized.

Text B

Malicious Software: Worms, Trojans and Bots

Dealing with malicious software, better known as malware, is a reality that we all face any time we connect to the Internet. Nobody wants to open up their e-mail to discover that they've just sent an infected file to all their friends, or that their data has been wiped by a virus. Although most people fear viruses, they are also surprisingly unaware of just what is out there in terms of malware and how it does its devious work. Here we'll look at some basic classes of malware and how they work to make your life miserable.

1. Malware Basics

Before we dig too deep into classes and types, we need to have a clear understanding of malware. Malware actually goes by another name, malicious code (or malcode). The "malicious" or "mal" (from the Latin "mallus," meaning "bad") ware means to attack, destroy, alter or otherwise damage the host machine on which it runs or the network to which that machine is attached. So, in short, malcode is dangerous code, and malware is dangerous software.

Although some malware can get into a machine through weaknesses in an operating system or a browser, most malware require a user to download it or somehow activate it by clicking a link or opening a file. Once the malware is active within a system, it will execute the instructions contained in its code.

There is no doubt that malware can do a lot of damage, such as changing how other applications work and locking or destroying data, but it does have limitations. Like legitimate software, malware cannot make any changes to a device's hardware. This means that even in the very worst-case scenario, a user can lose all of his or her data, but still recover the device by clearing it and reinstalling the operating system and other applications.

However, it is still best to avoid malware altogether. Becoming aware of the types of threats out there is one way that computer users can protect themselves.

2. Computer Viruses: Catching a Flu from Files

Viruses are probably the best-known type of malware. Like viruses in the natural world, computer viruses have two main purposes: to copy themselves and to spread. The actual damage a virus does depends on its designer. It is possible to have a benign virus that spreads without doing anything of note to the machines it infects.

Unfortunately, most viruses get into other programs, scripts[1] and other sets of instructions that are running on a device, and make changes in these areas. It is in this way that viruses destroy data, shut down programs and even prevent a computer from booting up.

3. Worms: Burrowing Their Way Through Your Network

Worms are very similar to viruses in that they are mostly concerned with copying themselves and spreading, but they use a different delivery system. Instead of spreading via infected files, worms use network vulnerabilities to travel from one host to another. This means that worms don't require a user to open anything or activate them in anyway. They just crawl in through a gap in a user's network security.

Once it gains access to a network, a worm looks for the next place to spread. While moving through the hosts in a network, the worm can do the same types of damage as a virus. Most worms also carry a payload[2], which is essentially a computer virus that the worm delivers once it reaches a new host. For example, the Blaster Worm, which appeared in 2003, carried a virus that caused computers running Windows to reboot multiple times. However, even seemingly harmless payload-free worms can overload a network and create a denial-of-service attack.

4. Trojans: More Interested in Controlling Your Computer Than Kidnapping Helen of Troy

Like the fabled wooden horse that was used to fool the people of Troy into letting the Greeks in, malware Trojans allow other people to gain access to your devices. Like a virus or a worm, a Trojan can run code that will damage or otherwise alter a device and its data. However, most Trojans are designed to open a back door into a system that a hacker can use to control and manipulate the device.

Unlike viruses and worms, Trojans don't copy themselves or attempt to spread to multiple computers. They are generally contained in a disguised file that depends on the user to activate it.

5. Bots: When Robots Rule the World

Bots are automated programs that carry out a specific process. There are many legitimate bots that help the Internet run smoothly, such as the Googlebot. However, bots can also be used to carry out more dubious processes, such as infecting unprotected computers and adding them to a malicious bot network (botnet[3]).

By remotely controlling a number of computers, the individual running the botnet can carry out many different types of attacks. For example, bots can steal data from the infected computer, including the user's contacts, passwords and other private information. Computers infected by bots may also become nodes for spreading spam, malware and other nasty surprises to other users. And finally, bots can use the infected network to launch denial-of-service attacks and other large-scale attacks. Bots are perhaps the most powerful type of malware in that they can be spread in many different ways and can attack using multiple methods.

6. Spyware: I Am Looking at You Right Now

Spyware doesn't attack your computer, but it still fits the definition of malware. Spyware collects information from your computer and sends it back to the program's creator, presumably so he or she can log in to your bank account or sell your personal information. Spyware is most often disguised as a free program to carry out another function, or it may be packaged with a legitimate piece of software.

7. How to Deal with It All: Common Sense Goes a Long Way

So now that you know about all these threats, how do you protect yourself?

The simple answer is that a bit of education and common sense is the most important aspect. It's pretty simple: Don't open e-mail attachments from people you don't know, and don't click on links from strangers. The limitation on viruses is that they have to be spread through infected files. In the vast majority of cases, a user must open the file to activate the virus.

The second thing you can do is always have up-to-date anti-virus software on your computer. The term "anti-virus" is getting somewhat dated. Most packages will protect you not just against viruses, but also other threats like worms and Trojans, but also spyware. There are many options out there both free and paid, which will give you solid protection from the vast majority of threats.

Finally, keeping your OS and your anti-virus system up to date is often enough to keep the malware out. Companies like Microsoft that make operating systems work very hard to keep on top of any new threats. You might not notice anything different while using a PC after a Windows update, but know that underneath the hood, there are significant updates that serve to plug any newly discovered security holes.

8. Conclusion

Malware isn't going away. In fact, as the number of people using Internet-enabled devices increases, the number and varieties of malware will likely increase as well. Being aware of the malware that is out there is the first step toward defending yourself from attacks. Most malware can be avoided by applying some commonsense when you are downloading and opening files from various sources. However, for a more complete sense of security, a trusted anti-virus program and a proper firewall can't be beat.

New Words

Bot	*n.* 自动运行型木马
infected	*adj.* 被感染的
wipe	*v.* 擦，揩，擦去
unaware	*adj.* 不知道的，没觉察到的
miserable	*adj.* 痛苦的，悲惨的，可怜的
destroy	*v.* 破坏，毁坏，消灭
browser	*n.* 浏览器
download	*n. & v.* 下载
instruction	*n.* 指令
limitation	*n.* 限制，局限性
recover	*v.* 重新获得，恢复
reinstall	*v.* 重新安装，重新设置

flu	*n.*	流感
spread	*v.*	传播，散布
crawl	*v.*	爬行，蠕动
gap	*n.*	缺口，裂口，间隙
payload	*n.*	有效数据，有效载荷
kidnap	*v.*	诱拐(小孩)，绑架，勒赎
fabled	*adj.*	寓言中的，虚构的
automate	*v.*	使自动化，自动操作
unprotected	*adj.*	无保护(者)，无防卫的
botnet	*n.*	僵尸网络
nasty	*adj.*	令人厌恶的，威胁的
presumably	*adv.*	推测起来，大概
stranger	*n.*	陌生人
up-to-date	*adj.*	直到现在的，最近的
variety	*n.*	品种，种类，变化，多样性
commonsense	*adj.*	具有常识的
beat	*v.*	打，打败

Phrases

malicious software	恶意软件
in terms of	根据，按照，用……的话，在……方面
malicious code(malcode)	恶意代码
host machine	主机
operating system(OS)	操作系统
no doubt	无疑地
become aware of	知道
benign virus	良性病毒
shut down	(使)机器等关闭
boot up	启动，引导
be similar to…	与……相似
be interested in…	对……有兴趣
in disguise	伪装，化装
in the majority	占多数，拥有多数

Notes

[1] Scripts are lists of commands executed by certain programs or scripting engines. They are usually text documents with instructions written using a scripting language. They are used to generate Web pages and to automate computer processes.

[2] A payload refers to the component of a computer virus that executes a malicious activity. Apart from the speed in which a virus spreads, the threat level of a virus is calculated by the damages it causes. Viruses with more powerful payloads tend to be more harmful.

Although not all viruses carry a payload, a few payloads are considered extremely dangerous. Some of the examples of payloads are data destruction, offensive messages and the delivery of spam emails through the infected user's account.

[3] A botnet is a group of computers connected in a coordinated fashion for malicious purposes. Each computer in a botnet is called a bot. These bots form a network of compromised computers, which is controlled by a third party and used to transmit malware or spam, or to launch attacks.

Exercises

[Ex. 5] Fill in the following blanks according to the text.

1. Although most people fear viruses, they are also surprisingly unaware of just what is out there ____________________ and ____________________.

2. Although some malware can get into a machine through ____________ in an operating system or a browser, most malware require a user to ____________ or somehow activate it by ________________or ________________.

3. There is no doubt that malware can do a lot of damage, such as ____________________ and ____________________, but it does have limitations. Malware cannot make any changes to ________________.

4. Viruses are probably the best-known type of _____________. Like viruses in the natural world, computer viruses have two main purposes: ________________ and ________________.

5. Worms are very _____________ viruses in that they are mostly concerned with copying themselves and spreading, but they use ______________________. Instead of spreading via infected files, worms use ____________________ to travel from one host to another.

6. Unlike viruses and worms, Trojans ____________________ or attempt to spread to ____________________. They are generally contained __________________ that depends on the user to activate it.

7. Bots are ______________ that carry out a specific process. Bots are perhaps the most powerful type of malware in that they can be spread ______________and can attack using ______________.

8. Spyware collects information from your computer and ________________________, presumably so he or she can ________________________ or sell your personal information. Spyware is most often disguised as ______________ to carry out another function, or it may be packaged with ________________________.

9. The first thing you should do to protect yourself against all the threats is ______________ from people you don't know, and __________________. The second thing you can do is always have ______________________ on your computer. Finally, ______________________ and

______________________ is often enough to keep the malware out.

10. According to the author, malware ____________________. In fact, as the number of people ____________________ increases, the number and ____________________ will likely increase as well.

Reading Material

Denial-of-Service Attack Techniques

1. Application-Layer Floods

Various DoS-causing exploits such as buffer[1] overflow can cause server-running software to get confused and fill the disk space or consume[2] all available memory or CPU time.

Other kinds of DoS rely primarily on brute force, flooding the target with an overwhelming[3] flux of packets, over saturating[4] its connection bandwidth or depleting the target's system resources. Bandwidth-saturating floods rely on the attacker having higher bandwidth available than the victim. A common way of achieving this today is via distributed denial-of-service, employing a botnet. Another target of DDoS attacks may be to produce added costs for the application operator, when the latter uses resources based on Cloud Computing. In this case normally application used resources are tied to[5] a needed Quality of Service (QoS) level (e.g. responses should be less than 200 ms) and this rule is usually linked to automated software (e.g. Amazon CloudWatch) to raise more virtual resources from the provider in order to meet the defined QoS levels for the increased requests. The main incentive[6] behind such attacks may be to drive the application owner to raise the elasticity levels in order to handle the increased application traffic, in order to cause financial losses or force them to become less competitive. Other floods may use specific packet types or connection requests to saturate finite[7] resources by, for example, occupying[8] the maximum number of open connections or filling the victim's disk space with logs.

A "banana attack" is another particular type of DoS. It involves redirecting[9] outgoing messages from the client back onto the client, preventing outside access, as well as flooding the client with the sent packets. A LAND[10] attack is of this type.

---

1 buffer *n.* 缓冲区

2 consume *v.* 消耗

3 overwhelming *adj.* 压倒性的，无法抵抗的

4 saturate *v.* 使饱和，浸透，使充满

5 tie to 依靠，依赖

6 incentive *n.* 动机 *adj.* 激励的

7 finite *adj.* 有限的

8 occupy *v.* 占用，占领，占据

9 redirect *v.* 使改道，使改变方向

10 LAND 局域网拒绝服务(Local Area Network Denial)

An attacker with shell-level access to a victim's computer may slow it until it is unusable or crash it by using a fork bomb.

A kind of application-level DoS attack is XDoS (or XML DoS) which can be controlled by modern web application firewalls (WAFs).

2. Degradation[1]-of-Service Attacks

"Pulsing" zombies[2] are compromised computers that are directed to launch intermittent and short-lived floodings of victim websites with the intent of merely slowing it rather than crashing it. This type of attack, referred to as "degradation-of-service" rather than "denial-of-service", can be more difficult to detect than regular zombie invasions[3] and can disrupt[4] and hamper[5] connection to websites for prolonged[6] periods of time, potentially causing more disruption than concentrated floods. Exposure of degradation-of-service attacks is complicated further by the matter of discerning whether the server is really being attacked or under normal traffic loads.

3. Denial-of-Service Level II

The goal of DoS L2 (possibly DDoS) attack is to cause a launching of a defense mechanism which blocks the network segment from which the attack originated. In case of distributed attack or IP header modification[7] (that depends on the kind of security behavior) it will fully block the attacked network from the Internet, but without system crash.

4. Distributed DoS Attack

A distributed[8] Denial-of-Service (DDoS) attack occurs when multiple systems flood the bandwidth or resources of a targeted system, usually one or more web servers. Such an attack is often the result of multiple compromised systems (for example, a botnet) flooding the targeted system with traffic. A botnet is a network of zombie computers programmed to receive commands without the owners' knowledge. When a server is overloaded with connections, new connections can no longer be accepted. The major advantages to an attacker of using a distributed denial-of-service attack are that multiple machines can generate more attack traffic than one machine, multiple attack machines are harder to turn off than one attack machine, and that the behavior of each attack machine can be stealthier[9], making it harder to track and shut down. These attacker advantages cause challenges for defense mechanisms. For example, merely purchasing more incoming bandwidth than the current volume of the attack might not help, because the attacker might be able to simply add more attack machines. This, after all, will end up completely crashing a website for periods of time.

---

1 degradation *n.* 降级，降格；退化

2 zombie *n.* 僵尸

3 invasion *n.* 入侵

4 disrupt *v.* 使中断

5 hamper *v.* 妨碍，牵制

6 prolonged *adj.* 延长的，拖延的

7 modification *n.* 更改，修改

8 distributed *adj.* 分布式的

9 stealthy *adj.* 鬼鬼祟祟的，秘密的，掩人耳目的

Malware can carry DDoS attack mechanisms; one of the better-known examples of this was MyDoom. Its DoS mechanism was triggered[1] on a specific date and time. This type of DDoS involved hardcoding the target IP address prior to release of the malware and no further interaction was necessary to launch the attack.

A system may also be compromised with a trojan, allowing the attacker to download a zombie agent[2], or the trojan may contain one. Attackers can also break into systems using automated tools that exploit flaws in programs that listen for connections from remote hosts. This scenario primarily concerns systems acting as servers on the web. Stacheldraht is a classic example of a DDoS tool. It utilizes a layered structure[3] where the attacker uses a client program to connect to handlers, which are compromised systems that issue commands to the zombie agents, which in turn facilitate the DDoS attack. Agents are compromised via the handlers by the attacker, using automated routines[4] to exploit vulnerabilities in programs that accept remote connections running on the targeted remote hosts. Each handler can control up to a thousand agents. In some cases a machine may become part of a DDoS attack with the owner's consent, for example, in Operation Payback, organized by the group Anonymous[5]. These attacks can use different types of Internet packets such as: TCP, UDP, ICMP etc.

These collections of systems compromisers are known as botnets / rootservers. DDoS tools like Stacheldraht still use classic DoS attack methods centered on IP spoofing and amplification[6] like smurf attacks and fraggle attacks (these are also known as bandwidth consumption attacks). SYN floods (also known as resource starvation attacks) may also be used. Newer tools can use DNS servers for DoS purposes. Unlike MyDoom's DDoS mechanism, botnets can be turned against any IP address. Script kiddies use them to deny the availability of well known websites to legitimate users. More sophisticated attackers use DDoS tools for the purposes of extortion[7]—even against their business rivals[8].

Simple attacks such as SYN floods may appear with a wide range of source IP addresses, giving the appearance of a well distributed DoS. These flood attacks do not require completion of the TCP three way handshake[9] and attempt to exhaust[10] the destination SYN queue[11] or the server bandwidth. Because the source IP addresses can be trivially spoofed, an attack could come from a limited set of sources, or may even originate from a single host. Stack[12] enhancements such as syn cookies may be effective mitigation against SYN queue flooding, however complete bandwidth

---

1 trigger *v.* 引发，引起，触发
2 agent *n.* 代理
3 layered structure 分层结构，成层结构
4 routine *n.* 例程，程序
5 anonymous *adj.* 匿名的
6 amplification *n.* 扩大
7 extortion *n.* 勒索，敲诈，强取
8 rival *n.* 竞争者，对手
9 handshake *n.* 握手
10 exhaust *v.* 用尽，耗尽
11 queue *n.* 队列
12 stack *v.* 堆叠

exhaustion[1] may require involvement.

If an attacker mounts an attack from a single host it would be classified as a DoS attack. In fact, any attack against availability would be classed as a denial-of-service attack. On the other hand, if an attacker uses many systems to simultaneously launch attacks against a remote host, this would be classified as a DDoS attack.

It has been reported that there are new attacks from Internet of things[2] which have been involved in denial of service attacks. In one noted attack that was made peaked at around 20,000 requests per second which came from around 900 CCTV cameras.

5. DDoS Extortion

In 2015, DDoS botnets such as DD4BC grew in prominence[3], taking aim at financial institutions. Cyber-extortionists typically begin with a low-level attack and a warning that a larger attack will be carried out if a ransom is not paid in Bitcoin[4]. Security experts recommend targeted websites to not pay the ransom[5]. The attackers tend to get into an extended extortion scheme once they recognize that the target is ready to pay.

6. Peer-to-Peer Attacks

Attackers have found a way to exploit a number of bugs in peer-to-peer servers to initiate DDoS attacks. The most aggressive of these peer-to-peer-DDoS attacks exploits DC++[6]. With peer-to-peer there is no botnet and the attacker does not have to communicate with the clients it subverts[7]. Instead, the attacker acts as a "puppet[8] master," instructing clients of large peer-to-peer file sharing hubs to disconnect[9] from their peer-to-peer network and to connect to the victim's website instead.

7. Permanent[10] Denial-of-Service Attack

Permanent denial-of-service (PDoS), also known loosely as phlashing, is an attack that damages a system so badly that it requires replacement or reinstallation of hardware. Unlike the distributed denial-of-service attack, a PDoS attack exploits security flaws which allow remote administration on the management interfaces[11] of the victim's hardware, such as routers, printers, or other networking hardware. The attacker uses these vulnerabilities to replace a device's firmware[12] with a modified, corrupt, or defective firmware image—a process which when done legitimately is known as flashing[13]. This therefore "bricks" the device, rendering it unusable for its original purpose

---

1 exhaustion *n.* 耗尽枯竭

2 Internet of things 物联网

3 prominence *n.* 突出，显著

4 Bitcoin *n.* 比特币

5 ransom *n.* 敲诈，勒索

6 DC++ is a free and open-source, peer-to-peer file-sharing client that can be used to connect to the Direct Connect network or to the ADC protocol.

7 subvert *v.* 推翻；暗中破坏，搅乱

8 puppet *n.* 傀儡，木偶

9 disconnect *v.* 断开，分离，拆开

10 permanent *adj.* 永久的，持久的

11 interface *n.* 接口

12 firmware *n.* 固件，韧件

13 flashing *n.* 刷新

until it can be repaired or replaced.

The PDoS is a pure hardware targeted attack which can be much faster and requires fewer resources than using a botnet or a root/vserver in a DDoS attack. Because of these features, and the potential and high probability of security exploits on Network Enabled Embedded Devices (NEEDs), this technique has come to the attention[1] of numerous hacking communities.

参 考 译 文

安 全 黑 客

安全黑客是一个试图破坏防御和利用计算机系统或网络弱点的人。黑客可能受到很多原因的驱动，如利益、收集信息、挑战及娱乐等，他们也会评估系统的弱点以协助防御潜在黑客。

1. 分类

计算机界持不同态度的几个地下小组使用不同的术语来划分自己，或尝试排除一些他们不认同的特定分组形式。

《新黑客词典》的作者埃里克·S. 雷蒙德主张把计算机地下组织的成员称为骇客。然而，这些人认为自己是黑客，甚至试图将雷蒙德的观点纳入他们所认为的更广泛的黑客文化中，雷蒙德已经严厉拒绝了这一观点。这些成员不是黑客/骇客，而是被分为一系列不同的类别，例如白帽、灰帽、黑帽和脚本小子。与雷蒙德的观点相反，他们通常认为进行更恶意活动的人是骇客。

根据拉尔夫·克里福德的观点，骇客是为了“擅自使用计算机进行犯罪，例如破坏该系统中的信息”。这些小组也可以由其活动的法律效应来界定。

（1）白帽

白帽黑客并非要恶意破坏安全性，他要么测试自己的安全系统，对客户进行渗透测试或漏洞评估，要么是为生产安全软件的安全公司工作。该术语通常是道德黑客的代名词，欧盟（还有其他机构）已经开发了认证、课件、课程和在线培训，涵盖了道德黑客的多个领域。

（2）黑帽

黑帽黑客是恶意或者以获取个人利益为目而破坏计算机安全的黑客。这个术语由理查德·斯托曼创造，将黑客的恶意与黑客文化中的玩耍和探索精神进行对比，或者与白帽黑客的精神进行对比。白帽黑客进行破解以找出需要修复的地方，也就意味着这是一种合法的职业行为。在流行文化中，黑帽黑客经常被描绘成典型的非法黑客团体，并且是“公众对计算机犯罪分子恐惧的化身”。

（3）灰帽

灰帽黑客介于黑帽和白帽黑客之间。例如，一个灰帽黑客可能会上网冲浪，进入一个计算机系统，唯一的目的是通知管理员他们的系统有一个安全缺陷。灰帽黑客可能会纠正该缺陷，但要收费。灰帽黑客有时会发现系统的缺陷，并将事实公开发布而不是告诉一群人。即

1 attention　*n.* 注意，关心，关注

使灰帽黑客可能不一定为了个人利益而进行黑客攻击，但未经授权的系统访问可以被视为非法的和不道德的。

（4）精英黑客

精英黑客在黑客中有社会地位，用精英黑客形容最具技巧的黑客。在这些黑客之间流传新发现的漏洞。如“欺骗大师”这样的精英团体对其成员赋予了一种信誉。

（5）脚本小子

脚本小子（也称为 skid 或 skiddie）是一个不熟练的黑客，通过使用其他人（通常由其他黑帽黑客）编写的自动化工具攻入计算机系统，因此脚本（即预先安排的计划或一组活动）+小孩（即小孩、孩子——缺乏知识和经验不成熟的个体）就构成了术语“脚本小子”，这种黑客通常对基本概念的理解甚微。

（6）蓝帽

蓝帽黑客是计算机安全咨询公司以外的人，他们在系统发布之前对其进行错误测试，寻找漏洞，以便关闭这些漏洞。Microsoft 还使用术语 BlueHat 代表一系列安全简报事件。

（7）黑客主义者

黑客主义者是利用技术宣传社会、思想、宗教或政治信息的黑客。

黑客主义可以分为两大类：

1）网络恐怖主义：涉及网站涂改或拒绝服务攻击的活动。

2）信息自由：让不公开的或以非机器可读格式存储的信息能够被公开访问。

2. 攻击

攻击互联网连接系统的典型方法是：

1）网络枚举：发现有关目标的信息。

2）漏洞分析：识别潜在的攻击方式。

3）利用：利用漏洞分析发现的漏洞来破坏系统。

为了做到这一点，计算机犯罪分子和安全专家经常使用几种工具和技术。

（1）安全漏洞利用

安全漏洞利用是针对已知弱点的一个现成的应用程序。安全漏洞利用的常见例子包括 SQL 注入、跨站点脚本和跨站点请求伪造，它们滥用不合规范编程实践可能引起的安全漏洞。其他漏洞利用可以通过文件传输协议（FTP）、超文本传输协议（HTTP）、PHP、SSH、Telnet 和一些网页来使用。这些在网站和网域黑客中很常见。

（2）技术

1）漏洞扫描器。漏洞扫描器是一种用于快速检查网络上的计算机以获取已知缺点的工具。黑客也常常使用端口扫描器。这些端口扫描器检查并查看指定计算机上的哪些端口是“打开”的或哪些是可以访问计算机的，有时会检测哪个程序或服务正在侦听该端口及其版本号。（防火墙通过限制对端口和计算机的访问来保护计算机免受入侵者的攻击，但仍可以绕过它们。）

2）查找漏洞。黑客也可能会尝试手动查找漏洞。通常的方法是在计算机系统的代码中搜索可能的漏洞，然后测试它们，有时如果没有提供代码，则会采用逆向工程方法。

3）暴力攻击。密码猜测这种方法在用于检查所有短密码时非常快，但是对于较长的密码，则会使用诸如字典攻击等其他方法，因为强力搜索颇为费时。

4）密码破解。密码破解是从计算机系统存储或传输的数据中提取密码的过程。常见的方法包括重复尝试猜测密码、手工尝试最常用的密码以及不断尝试使用“字典”中的密码或具有多个密码的文本文件。

5）数据包分析器。数据包分析器（“数据包嗅探器”）是捕获数据包的应用程序，可用于捕获通过网络传输的密码和其他数据。

6）欺骗攻击（phishing）。欺骗攻击指一个程序、系统或网站通过伪造数据成功地伪装成另一个程序、系统或网站，从而被用户或其他程序视为可信赖的系统——通常欺骗程序、系统或用户来显示机密信息，例如用户名和密码。

7）Rootkit。Rootkit 是一个使用低级、难以检测的方法来破坏合法操作员对操作系统控制的程序。它通常会掩盖其安装，并尝试通过破坏标准系统安全性来防止被删除。它们可能替换系统的二进制文件，这样检查进程表几乎不可能发现它们。

8）特洛伊木马。特洛伊木马是一个程序，似乎在做一件事，但实际上正在做另一件事。它可用于在计算机系统中设置后门，以便入侵者以后能够访问。（这个名字是指特洛伊战争中的木马，概念上类似欺骗捍卫者将入侵者带入保护区。）

9）计算机病毒。病毒是一个自我复制的程序，通过将自身的副本插入其他可执行代码或文档来传播。这样做的行为类似于生物病毒，即通过将自身插入活细胞而传播。虽然一些病毒是无害的或只是恶作剧，但大多数被认为是恶意的。

10）计算机蠕虫。像病毒一样，蠕虫也是一个自我复制的程序。它不同于病毒，它通过计算机网络传播而无须用户干预，也不需要附加到现有的程序。尽管如此，许多人不加区别地使用术语“病毒”和“蠕虫”来描述任何自我传播的程序。

11）键盘记录器。键盘记录器是一种工具，用于在受影响的计算机上记录（“记日志”）每个按键以备以后检索。通常允许该工具的使用者访问在受影响的计算机上键入的机密信息。一些键盘记录器使用病毒、木马和类似 Rootkit 的方法来隐藏自身。然而，其中一些用于合法目的，甚至用于增强计算机的安全性。例如，企业可以在销售点使用的计算机上安装键盘记录器，以发现雇员欺诈的证据。

Unit 5

Text A

Firewall

A firewall is a part of a computer system or network that is designed to block unauthorized access while permitting authorized communications. It is a device or set of devices which is configured to permit or deny computer applications based upon a set of rules and other criteria (see Figure 5-1).

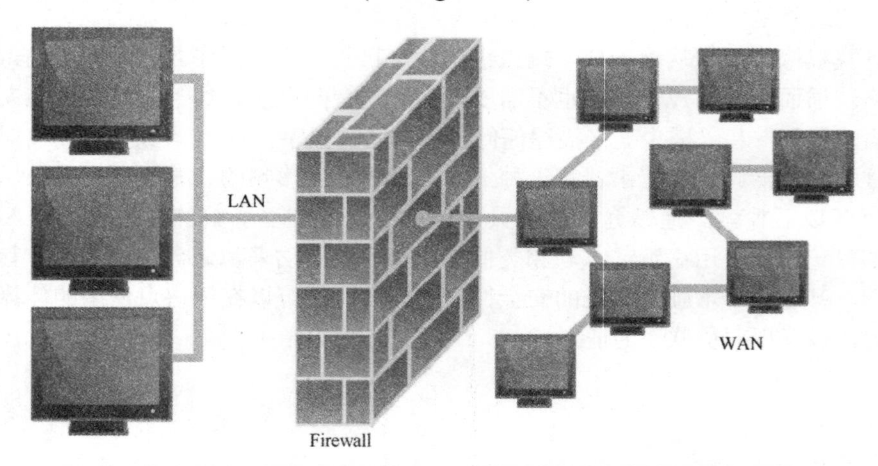

Figure 5-1 An Illustration of Where a Firewall Would Be Located in a Network

Firewalls can be implemented in either hardware or software, or a combination of both. Firewalls are frequently used to prevent unauthorized Internet users from accessing private networks connected to the Internet, especially Intranets. All messages entering or leaving the Intranet pass through the firewall, which examines each message and blocks those that do not meet the specified security criteria.

There are several types of firewall techniques:

1) Packet filter: Packet filtering inspects each packet passing through the network and accepts or rejects it based on user-defined rules. Although difficult to configure, it is fairly effective and mostly transparent to its users. It is susceptible to IP spoofing.

2) Application gateway: Applies security mechanisms to specific applications, such as FTP and

Telnet servers. This is very effective, but can impose performance degradation.

3) Circuit-level gateway[1]: Applies security mechanisms when a TCP or UDP connection is established. Once the connection has been made, packets can flow between the hosts without further checking.

4) Proxy server: Intercepts all messages entering and leaving the network. The proxy server effectively hides the true network addresses.

1. Function

A firewall is a dedicated appliance, or software running on a computer, which inspects network traffic passing through it, and denies or permits passage based on a set of rules/criteria.

It is normally placed between a protected network and an unprotected network and acts like a gate to protect assets to ensure that nothing private goes out and nothing malicious comes in.

A firewall's basic task is to regulate some of the flow of traffic between computer networks of different trust levels. Typical examples are the Internet which is a zone with no trust and an Internal network which is a zone of higher trust. A zone with an intermediate trust level, situated between the Internet and a trusted internal network, is often referred to as a "perimeter network" or Demilitarized zone (DMZ).

A firewall's function within a network is similar to physical firewalls with fire doors in building construction. In the network, it is used to prevent network intrusion to the private network. In the building, it is intended to contain and delay structural fire from spreading to adjacent structures.

2. Types

There are several classifications of firewalls depending on where the communication is taking place, where the communication is intercepted and the state that is being traced.

(1) Network Layer

Network layer firewalls, also called packet filters, operate at a relatively low level of the TCP/IP protocol stack, not allowing packets to pass through the firewall unless they match the established rule set. The firewall administrator may define the rules; or default rules may apply. The term "packet filter" originated in the context of BSD operating systems.

Network layer firewalls generally fall into two sub-categories, stateful and stateless. Stateful firewalls[2] maintain context about active sessions, and use that "state information" to speed packet processing. Any existing network connection can be described by several properties, including source and destination IP address, UDP or TCP ports, and the current stage of the connection's lifetime (including session initiation, handshaking, data transfer, or completion connection). If a packet does not match an existing connection, it will be evaluated according to the rule set for new connections. If a packet matches an existing connection based on comparison with the firewall's state table, it will be allowed to pass without further processing.

Stateless firewalls require less memory, and can be faster for simple filters that require less time to filter than to look up a session. They may also be necessary for filtering stateless network protocols that have no concept of a session. However, they cannot make more complex decisions based on what stage communications between hosts have reached.

Modern firewalls can filter traffic based on many packet attributes like source IP address, source port, destination IP address or port, destination service like WWW or FTP. They can filter based on protocols, TTL values, netblock[3] of originator, of the source, and many other attributes.

(2) Application-Layer

Application-layer firewalls work on the application level of the TCP/IP stack (i.e., all browser traffic, or all telnet or ftp traffic), and may intercept all packets traveling to or from an application. They block other packets (usually dropping them without acknowledgment to the sender). In principle, application firewalls can prevent all unwanted outside traffic from reaching protected machines.

On inspecting all packets for improper content, firewalls can restrict or prevent outright the spread of networked computer worms and Trojan. The additional inspection criteria can add extra latency to the forwarding of packets to their destination.

(3) Proxies

A proxy device (running either on dedicated hardware or as software on a general-purpose machine) may act as a firewall by responding to input packets (connection requests, for example) in the manner of an application, whilst blocking other packets.

Proxies make tampering with an internal system from the external network more difficult and misuse of one internal system would not necessarily cause a security breach exploitable from outside the firewall (as long as the application proxy remains intact and properly configured). Conversely, intruders may hijack a publicly-reachable system and use it as a proxy for their own purposes; the proxy then masquerades as that system to other internal machines. While use of internal address spaces enhances security, crackers may still employ methods such as IP spoofing to attempt to pass packets to a target network.

(4) Network Address Translation

Firewalls often have Network Address Translation (NAT) functionality, and the hosts protected behind a firewall commonly have addresses in the "private address range", as defined in RFC 1918. Firewalls often have such functionality to hide the true address of protected hosts. Originally, the NAT function was developed to address the limited number of IPv4 routable addresses that could be used or assigned to companies or individuals as well as to reduce both the amount and cost of obtaining enough public addresses for every computer in an organization. Hiding the addresses of protected devices has become an increasingly important defense against network reconnaissance.

New Words

deny	*v.*	否认，拒绝
criteria	*n.*	标准
combination	*n.*	结合，联合，合并
inspect	*v.*	检查
reject	*v.*	拒绝，抵制，否决，丢弃
transparent	*adj.*	透明的，显然的，明晰的
spoof	*v.*	哄骗，欺骗
intercept	*v.*	中途阻止，截取

hide *v.* 隐藏，掩藏，隐瞒
intermediate *adj.* 中间的
n. 媒介
situated *adj.* 位于……的，处于……境遇的
originate *v.* 起源，发生
stateless *adj.* 无状态的
property *n.* 性质，特性
stage *n.* 发展的进程、阶段或时期
handshake *n.* 握手
acknowledgment *n.* 承认
sender *n.* 寄件人，发送人
principle *n.* 法则，原则，原理
worm *n.* 蠕虫
latency *n.* 反应时间，潜伏，潜在，潜伏物
tamper *v.* 篡改
misuse *v.&n.* 误用，错用，滥用
breach *n.* 破坏，裂口
v. 打破，突破
exploitable *adj.* 可开发的，可利用的
intact *adj.* 完整无缺的
hijack *v.* 劫持
masquerade *v.* 化装；伪装
cracker *n.* 骇客，解密高手
increasingly *adv.* 日益，愈加
reconnaissance *n.* 侦察，搜索

Phrases

base upon 根据，依据
pass through 经过，通过
packet filter 包过滤
be susceptible to… 对……敏感，可被……
circuit-level gateway 电路级网关
perimeter network 外围网络
fire door 防火门
spread to 传到，波及，蔓延到
stateful firewall 状态防火墙
stateless firewall 无状态防火墙
take place 发生，进行
rule set 规则集

fall into　　分成，属于
active session　　有效对话期间，工作时间
state table　　状态表
routable address　　可路由的地址

Abbreviations

UDP (User Datagram Protocol)　　用户数据报协议
DMZ (DeMilitarized Zone)　　非军事区
BSD (Berkeley Software Distribution)　　伯克利软件发布
TTL (Time To Live)　　生存时间
NAT (Network Address Translation)　　网络地址转换

Notes

[1] Circuit-level gateways work at the session layer of the OSI model, or as a "shim-layer" between the application layer and the transport layer of the TCP/IP stack. They monitor TCP handshaking between packets to determine whether a requested session is legitimate. Information passed to a remote computer through a circuit-level gateway appears to have originated from the gateway. Circuit-level firewall applications represent the technology of next to first generation. Firewall technology supervises TCP handshaking among packets to confirm a session is genuine. Firewall traffic is clean based on particular session rules and may be controlled to acknowledged computers only. Circuit-level firewalls conceal the network itself from the external, which is helpful for interdicting access to impostors. But circuit-level firewalls do not clean entity packets. This is useful for hiding information about protected networks. Circuit-level gateways are relatively inexpensive and have the advantage of hiding information about the private network they protect. On the other hand, they do not filter individual packets.

[2] In computing, a stateful firewall is a network firewall that tracks the operating state and characteristics of network connections traversing it. The firewall is configured to distinguish legitimate packets for different types of connections. Only packets matching a known active connection are allowed to pass the firewall.

Stateful Packet Inspection (SPI), also referred to as dynamic packet filtering, is a security feature often included in business networks.

[3] A netblock is a range of IP addresses. ISPs must be assigned addresses in blocks, and the larger the block that can be assigned to someone the better because that makes for less entries in the Internet routing tables for information on how to get to those addresses.

Exercises

[Ex. 1]　Answer the following questions according to the text.

1. What is a firewall?

2. What can firewalls be implemented in?

3. What are the types of firewall techniques mentioned in the text?

4. What is a firewall's basic function?

5. What is a firewall's function within a network similar to?

6. What do network layer firewalls do? What did the term "packet filter" originate?

7. What are the two sub-categories network layer firewalls generally fall into? What do they do respectively?

8. Where do application-layer firewalls work? What may they do?

9. What may a proxy device (running either on dedicated hardware or as software on a general-purpose machine) act as?

10. What was the NAT function developed to do originally?

[Ex. 2] Translate the following terms or phrases from English into Chinese and vice versa.

1.	rule set	1.	
2.	packet filter	2.	
3.	state table	3.	
4.	active session	4.	
5.	routable address	5.	
6.	握手	6.	
7.	检查	7.	
8.	中途阻止，截取	8.	
9.	侦察，搜索	9.	
10.	篡改	10.	

[Ex. 3] Translate the following passage into Chinese.

What Is a Network Firewall?

A network firewall protects a computer network from unauthorized access. Network firewalls may be hardware devices, software programs, or a combination of the two.

Network firewalls guard an internal computer network, for example, a home, school, business, or intranet, against malicious access from the outside. They may also be configured to limit access to the outside from internal users.

1. Network Firewalls and Broadband Routers

Many home network router products include built-in firewall support.

The administrative interface of these routers includes configuration options for the firewall. Router firewalls can be turned off (disabled), or they can be set to filter certain types of network traffic through so-called firewall rules.

2. Network Firewalls and Proxy Servers

Another common form of network firewall is a proxy server. Proxy servers act as an intermediary between internal computers and external networks by receiving and selectively blocking data packets at the network boundary. These network firewalls also provide an extra measure of safety by hiding internal LAN addresses from the outside Internet. In a proxy server firewall environment, network requests from multiple clients appear to the outsider as all coming from the same proxy server address.

[Ex. 4] Fill in the blanks with the words given below.

specifically	denied	existing	passes	connections
efficient	trusted	emerged	packet	vulnerable

Firewall

A firewall is a network security system, either hardware- or software-based, that uses rules to control incoming and outgoing network traffic. It acts as a barrier between a ___1___ network and an untrusted network. It controls access to the resources of a network through a positive control model. This means that the only traffic allowed onto the network is defined in the firewall policy; all other traffic is ___2___.

1. Packet Firewalls

The earliest firewalls functioned as packet filters, inspecting the packets that are transferred between computers on the Internet. When a packet ___3___ through a packet filter firewall, its source and destination address, protocol, and destination port number are checked against the firewall's rule set. Any packets that aren't ___4___ allowed onto the network are dropped (i.e., not forwarded to their destination). For example, if a firewall is configured with a rule to block Telnet access, then the firewall will drop packets destined for TCP port number 23, the port where a Telnet server application would be listening.

Packet-filter firewalls work mainly on the first three layers of the OSI reference model (physical, data-link and network), although the transport layer is used to obtain the source and destination port numbers. While generally fast and ___5___, they have no ability to tell whether a packet is part of an existing stream of traffic. Because they treat each packet in isolation, this makes them ___6___ to spoofing attacks and also limits their ability to make more complex decisions based on what stage communications between hosts are at.

2. Stateful Firewalls

In order to recognize a packet's connection state, a firewall needs to record all connections passing through it to ensure it has enough information to assess whether a packet is the start of a new connection, a part of an ___7___ connection, or not part of any connection. This is what's called "stateful packet inspection". Stateful inspection was first introduced in 1994 by Check Point Software in its FireWall-1 software firewall, and by the late 1990s, it was a common firewall product feature.

This additional information can be used to grant or reject access based on the packet's history in the state table, and to speed up ___8___ processing; that way, packets that are part of an existing connection based on the firewall's state table can be allowed through without further analysis. If a packet does not match an existing connection, it's evaluated according to the rule set for new connections.

3. Application-Layer Firewalls

As attacks against Web servers became more common, so too did the need for a firewall that could protect servers and the applications running on them, not merely the network resources behind them. Application-layer firewall technology first ___9___ in 1999, enabling firewalls to inspect and filter packets on any OSI layer up to the application layer.

The key benefit of application-layer filtering is the ability to block specific content, such as known malware or certain websites, and recognize when certain applications and protocols—such as HTTP, FTP and DNS—are being misused.

4. Proxy Firewalls

Firewall proxy servers also operate at the firewall's application layer, acting as an intermediary for requests from one network to another for a specific network application. A proxy firewall prevents direct ___10___ between either sides of the firewall; both sides are forced to conduct the session through the proxy, which can block or allow traffic based on its rule set. A proxy service must be run for each type of Internet application the firewall will support, such as an HTTP proxy for Web services.

Text B

Antivirus Software

Antivirus or antivirus software (often abbreviated as AV), sometimes known as anti-malware software, is computer software used to prevent, detect and remove malicious software.

Antivirus software was originally developed to detect and remove computer viruses, hence the name. However, with the proliferation of other kinds of malware, antivirus software started to provide protection from other computer threats.

1. Identification Methods

One of the few solid theoretical results in the study of computer viruses is Frederick B. Cohen's 1987 demonstration that there is no algorithm that can perfectly detect all possible viruses. However, using different layers of defense, a good detection rate may be achieved.

There are some methods which antivirus engine can use to identify malware.

- Sandbox[1] detection. It is a particular behavioural-based detection technique that, instead of detecting the behavioural fingerprint at run time, it executes the programs in a virtual environment, logging what actions the program performs. Depending on the actions logged, the antivirus engine can determine if the program is malicious or not. If not, the program is executed in the real environment. Albeit this technique has shown to be quite effective, given its heaviness and slowness, it is rarely used in end-user antivirus solutions.
- Data mining[2] techniques. They are one of the latest approachs applied in malware detection. Data mining and machine learning algorithms are used to try to classify the behaviour of a file (as either malicious or benign) given a series of file features, which are extracted from the file itself.

(1) Signature-Based Detection

Traditional antivirus software relies heavily upon signatures to identify malware.

Substantially, when a malware arrives in the hands of an antivirus firm, it is analyzed by malware researchers or by dynamic analysis systems. Then, once it is determined to be a malware, a proper signature of the file is extracted and added to the signatures database of the antivirus software.

Although the signature-based approach can effectively contain malware outbreaks, malware authors have tried to stay a step ahead of such software by writing "oligomorphic", "polymorphic" and, more recently, "metamorphic[3]" viruses, which encrypt parts of themselves or otherwise modify themselves as a method of disguise, so as to not match virus signatures in the dictionary.

(2) Heuristics

Many viruses start as a single infection and through either mutation or refinements by other attackers, they can grow into dozens of slightly different strains, called variants. Generic detection refers to the detection and removal of multiple threats using a single virus definition.

For example, the Vundo trojan has several family members, depending on the antivirus vendor's classification. Symantec classifies members of the Vundo family into two distinct categories: Trojan Vundo and Trojan Vundo B.

While it may be advantageous to identify a specific virus, it can be quicker to detect a virus family through a generic signature or through an inexact match to an existing signature. Virus researchers find common areas that all viruses in a family share uniquely and can thus create a single generic signature. These signatures often contain non-contiguous code, using wildcard characters where differences lie. These wildcards allow the scanner to detect viruses even if they are padded with extra, meaningless code. A detection that uses this method is said to be "heuristic detection".

(3) Rootkit Detection

Antivirus software can attempt to scan for rootkits. A rootkit is a type of malware designed to gain administrative-level control over a computer system without being detected. Rootkits can change how the operating system functions and in some cases can tamper with the antivirus program and render it ineffective. Rootkits are also difficult to remove, in some cases requiring a complete reinstallation of the operating system.

(4) Real-Time Protection

Real-time protection, on-access scanning, background guard, resident shield, autoprotect, and other synonyms refer to the automatic protection provided by most antivirus, anti-spyware, and other anti-malware programs. This monitors computer systems for suspicious activity such as computer viruses, spyware, adware, and other malicious objects in "real-time", in other words while data loaded into the computer's active memory: When inserting a CD, opening an e-mail, or browsing the web, or when a file already on the computer is opened or executed.

2. Performance and Other Drawbacks

Antivirus software has some drawbacks, first of which is that it can impact a computer's performance. Furthermore, inexperienced users can be lulled into a false sense of security when using the computer, considering themselves to be invulnerable, and may have problems understanding the prompts and decisions that antivirus software presents them with. An incorrect decision may lead to a security breach. If the antivirus software employs heuristic detection, it must be fine-tuned to minimize misidentifying harmless software as malicious (false positive).

Antivirus software itself usually runs at the highly trusted kernel level of the operating system to allow it access to all the potential malicious process and files, creating a potential avenue of attack. The UK and US intelligence agencies, GCHQ and the National Security Agency (NSA), respectively, have been exploiting antivirus software to spy on users. Antivirus software has highly privileged and trusted access to the underlying operating system, which makes it a much more appealing target for remote attacks. Additionally antivirus software is "years behind security-conscious client-side applications like browsers or document readers", according to Joxean Koret, a researcher with Coseinc, and a Singapore-based information security consultancy.

3. Alternative Solutions

Installed antivirus solutions, running on individual computers, although the most used, is only one method of guarding against malware. Other alternative solutions are also used, including hardware and network firewalls, Cloud antivirus and online scanners.

(1) Hardware and Network Firewall

Hardware firewalls are mostly seen in broadband modems, and is the first line of defense, using packet filtering. Before an Internet packet reaches your PC, the hardware firewall will monitor the packets and check where it comes from. It also checks if the IP address or header can be trusted. After these checks, the packet then reaches your PC. It blocks any links that contain malicious behavior based on the current firewall setup in the device. A hardware firewall usually does not need a lot of configuration. Most of the rules are built-in and predefined and based on these inbuilt rules. Most hardware firewalls will have a minimum of four network ports to connect other computers, but for larger networks, business networking firewall solutions are available.

Network firewalls protect a computer network from unauthorized access. They may be hardware devices, software programs, or a combination of the two. They may protect against infection from outside the protected computer or network, and limit the activity of any malicious software which is present by blocking incoming or outgoing requests on certain TCP/IP ports.

The types of threats a firewall can protect against include:

Unauthorized access to network resources—an intruder may break into a host on the network and gain unauthorized access to files.

Denial of service—an individual from outside of the network could, for example, send thousands of mail messages to a host on the net in an attempt to fill available disk space or load the network links.

Masquerading—electronic mail appearing to have originated from one individual could have been forged by another with the intent to embarrass or cause harm.

(2) Cloud Antivirus

Cloud antivirus is a technology that uses lightweight agent software on the protected computer, while offloading the majority of data analysis to the provider's infrastructure.

One approach to implementing cloud antivirus involves scanning suspicious files using multiple antivirus engines. This approach was proposed by an early implementation of the cloud antivirus concept called CloudAV. CloudAV was designed to send programs or documents to a network cloud where multiple antivirus and behavioral detection programs are used simultaneously in order to improve detection rates. Parallel scanning of files using potentially incompatible antivirus scanners is achieved by spawning a virtual machine per detection engine and therefore eliminating any possible issues. CloudAV can also perform "retrospective detection", whereby the cloud detection engine rescans all files in its file access history when a new threat is identified, thus improving new threat detection speed. Finally, CloudAV is a solution for effective virus scanning on devices that lack the computing power to perform the scans themselves.

Some examples of cloud antivirus products are Panda Cloud Antivirus, Crowdstrike, Cb Defense and Immunet. Comodo group has also produced cloud-based antivirus.

(3) Online Scanning

Some antivirus vendors maintain websites with free online scanning capability of the entire computer, critical areas only, local disks, folders or files. Periodic online scanning is a good idea for those that run antivirus applications on their computers because those applications are frequently slow to catch threats. One of the first things that malicious software does in an attack is disable any existing antivirus software and sometimes the only way to know of an attack is by turning to an online resource that is not installed on the infected computer.

New Words

antivirus	*n.*	抗病毒软件
proliferation	*n.*	增殖，繁殖
protection	*n.*	保护
theoretical	*adj.*	理论的
demonstration	*n.*	示范，实证
sandbox	*n.*	沙箱，沙盒
virtual	*adj.*	虚拟的；实质的
perform	*v.*	履行，执行
heaviness	*n.*	沉重，重量
analyze	*v.*	分析，分解
researcher	*n.*	研究者，研究员
outbreak	*n.*	(战争的)爆发，(疾病的)发作
polymorphic	*adj.*	多形的，多态的
metamorphic	*adj.*	变形的，变质的，改变结构的
disguise	*v.*	假装，伪装，掩饰
	n.	伪装
heuristic	*adj.*	启发式的
infection	*n.*	传染，感染

mutation	*n.* 变化，转变，(生物物种的)突变
refinement	*n.* 改良品，细微的改良
variant	*n.* 变种，变量
	adj. 不同的，变异的
inexact	*adj.* 不精确的
wildcard	*n.* 通配符
meaningless	*adj.* 无意义的
code	*n.* 代码，编码
ineffective	*adj.* 无效的
protection	*n.* 保护
scanning	*v.* 扫描
background	*n.* 后台，背景
monitor	*v.* 监控
	n. 监视器，监控器
adware	*n.* 广告软件，恶意广告软件
drawback	*n.* 缺点，障碍
impact	*v.* 对……发生影响
	n. 影响，效果
performance	*n.* 性能，执行
inexperienced	*adj.* 无经验的，不熟练的
invulnerable	*adj.* 不会受伤害的
prompt	*v.* 提示
	n. 提示符，提示信息
misidentify	*v.* 错误识别，误认
kernel	*n.* 内核
avenue	*n.* 方法，途径
privileged	*adj.* 有特权的
block	*v.* 阻塞
configuration	*n.* 配置，构造，结构
predefine	*v.* 预先确定，预定义
inbuilt	*adj.* 嵌入的，内置的
port	*n.* 端口
intruder	*n.* 入侵者
embarrass	*v.* 使困窘，阻碍，麻烦
harm	*v.* & *n.* 伤害，损害
offload	*v.* 卸下，减轻（负担）
suspicious	*adj.* 可疑的，怀疑的
simultaneously	*adv.* 同时地
incompatible	*adj.* 不兼容的，矛盾的，不调和的

spawn	*v.* 大量产生，造成
retrospective	*adj.* 回顾的
rescan	*v.* & *n.* 重新扫描
periodic	*adj.* 周期的，定期的
frequently	*adv.* 常常，频繁地
catch	*n.* 捕捉
	v. 捕获
disable	*v.* 使残废，使失去能力，丧失能力

Phrases

antivirus software	抗病毒软件，反病毒软件
anti-malware software	抗恶意程序软件，反恶意程序软件
computer threat	计算机威胁
antivirus engine	抗病毒引擎，反病毒引擎
instead of…	代替，而不是……
virtual environment	虚拟环境
data mining	数据挖掘
a series of	一连串的，一系列的
arrive in	到达
dynamic analysis	动态分析
real-time protection	实时保护，实时防护
background guard	后台保护
resident shield	驻留防护
automatic protection	自动保护
National Security Agency	(美国)国家安全局
spy on	侦查，暗中监视
packet filtering	包过滤
in an attempt to	力图，试图
virtual machine	虚拟机

Abbreviations

CD (Compact Disc)	光盘
GCHQ (Government Communications Headquarters)	(美国)国家通信总局
TCP(Transmission Control Protocol)	传输控制协议

Notes

[1] In computer security, a sandbox is a security mechanism for separating running programs. It is often used to execute untested or untrusted programs or code, possibly from unverified or untrusted third parties, suppliers, users or websites, without risking harm to the host machine or

operating system. A sandbox typically provides a tightly controlled set of resources for guest programs to run in, such as scratch space on disk and memory. Network access, the ability to inspect the host system or read from input devices are usually disallowed or heavily restricted.

[2] Data mining is the computing process of discovering patterns in large data sets involving methods at the intersection of machine learning, statistics, and database systems. It is an interdisciplinary subfield of computer science. The overall goal of the data mining process is to extract information from a data set and transform it into an understandable structure for further use. Aside from the raw analysis step, it involves database and data management aspects, data pre-processing, model and inference considerations, interestingness metrics, complexity considerations, post-processing of discovered structures, visualization, and online updating.

[3] Metamorphic code is code that when run outputs a logically equivalent version of its own code under some interpretation. This is similar to a quine, except that a quine's source code is exactly equivalent to its own output. Metamorphic code also usually outputs machine code and not its own source code.

Exercises

[Ex. 5] Fill in the following blanks according to the text.

1. Antivirus or antivirus software (often abbreviated as AV), sometimes known as ________________, is computer software used to __________, _________ and remove _________________.

2. Although the signature-based approach can effectively contain __________________, malware authors have tried to stay a step ahead of such software by writing ______________, _______________ and, more recently, ________________ viruses, which encrypt parts of themselves or otherwise modify themselves as a method of disguise, so as to ______________ in the dictionary.

3. Many viruses start as ________________ and through either mutation or refinements by other attackers, they can grow into dozens of slightly different strains, called __________. Generic detection refers to the _____________ and removal of _____________ using a single virus definition.

4. A rootkit is a type of malware designed to ________________ over a computer system without being detected. Rootkits can change ________________ and in some cases can tamper with _________________ and render it ineffective. Rootkits are also difficult to remove, in some cases requiring _____________________ of the operating system.

5. Real-time protection monitors computer systems for suspicious activity such as computer viruses, ______________, ______________, and ______________ in "real-time", in other words while data loaded into the computer's active memory: When inserting a CD, ______________, or ______________, or when a file already on the computer is opened or executed.

6. Antivirus software has some drawbacks, first of which is that it can ______________________. Furthermore, inexperienced users can be lulled into a false sense of

security when using the computer, considering themselves ____________________, and may have problems understanding ____________________ that antivirus software presents them with.

7. Hardware firewalls are mostly seen in______________, and is ________________, using ______________. A hardware firewall usually ______________________. Most hardware firewalls will have __________________ network ports to connect other computers.

8. Network firewalls protect a computer network from unauthorized access. They may be ________________, __________________, or ______________________. The types of threats a firewall can protect against include unauthorized access to ______________, ________________, and _________________.

9. Cloud antivirus is a technology that uses _________________on the protected computer, while offloading the majority of data analysis to ________________. Some examples of cloud antivirus products are _________________ , _________________, Cb Defense and Immunet. Comodo group has also produced _________________.

10. One of the first things that malicious software does in an attack is ______________ and sometimes the only way to know of an attack is by turning to ______________________ that is not installed on ____________________.

Reading Material

Website Security

Websites are unfortunately prone to[1] security risks. And so are any networks to which web servers are connected. Setting aside[2] risks created by employee use or misuse of network resources, your web server and the site it hosts present your most serious sources of security risk.

Web servers by design open a window between your network and the world. The care taken with server maintenance, web application updates and your website coding will define the size of that window, limit the kind of information that can pass through it and thus establish[3] the degree of web security you will have.

1. Is Your Site or Network at Risk?

"Web security" is relative and has two components, one internal and one public. Your relative security is high if you have few network resources of financial value, your company and site aren't controversial[4] in any way, your network is set up with tight permissions, your web server is patched up[5] to date with all settings done correctly, your applications on the web server are all patched and updated, and your website code is done to high standards.

---

1 be prone to　倾向于，有……的倾向，容易

2 set aside　不顾，取消

3 establish　*v.*　建立，设立，安置

4 controversial　*adj.*　争论的，争议的

5 patch up　修补

Your web security is relatively lower if your company has financial assets like credit card or identity information, if your website content is controversial, your servers, applications and site code are complex or old and are maintained by an underfunded or outsourced IT department. All IT departments are budget[1] challenged and tight staffing often creates deferred maintenance issues that play into the hands of[2] any who want to challenge your web security.

2. Web Security Risk—Should You Be Worried?

If you have assets of importance or if anything about your site puts you in the public spotlight[3] then your web security will be tested. We hope that the information provided here will prevent you and your company from being embarrassed, or worse.

It's well known that poorly written software creates security issues. The number of bugs that could create web security issues is directly proportional[4] to the size and complexity[5] of your web applications and web server. Basically, all complex programs either have bugs or at the very, least weaknesses. On top of that, web servers are inherently[6] complex programs. Websites are themselves complex and intentionally invite ever greater interaction with the public. And so there are many and growing opportunities for security holes.

Technically, the very same programming that increases the value of a website, namely interaction with visitors, also allows scripts or SQL commands to be executed on your web and database servers in response to visitor requests. Any web-based form or script installed at your site may have weaknesses or outright bugs and every such issue presents a web security risk.

A web security issue is faced by site visitors as well. A common website attack involves the silent[7] and concealed[8] installation of code that will exploit the browsers of visitors. Your site is not the end target at all in these attacks. There are, at this time, many thousands of websites out there that have been compromised. The owners have no idea that anything has been added to their sites and that their visitors are at risk. In the meantime visitors are being subject to[9] attack and successful attacks are installing nasty code onto the visitor's computers.

3. Web Server Security

The world's most secure web server is the one that is turned off. Simple, bare-bones web servers that have few open ports and few services on those ports are the next best thing. This just isn't an option for most companies. Powerful and flexible applications are required to run complex sites and these are naturally more subject to web security issues.

Any system with multiple open ports, multiple services and multiple scripting languages is

---

1 budget *n.* 预算 *v.* 做预算，编入预算

2 play into the hands of 故意使……占便宜，为……谋方便

3 spotlight *n.* 聚光灯

4 proportional *adj.* 成比例的，相称的

5 complexity *n.* 复杂性

6 inherently *adv.* 天性地，固有地

7 silent *adj.* 寂静的，沉默的，无声的，无记载的

8 conceal *v.* 隐藏，隐蔽，隐瞒

9 be subject to… 常遭受……

vulnerable simply because it has so many points of entry to watch.

If your system has been correctly configured and your IT staff has been very punctual[1] about applying security patches and updates your risks are mitigated[2]. Then there is the matter of the applications you are running. These too require frequent updates. And last there is the website code itself.

4. WebSite Code and Web Security

Your site undoubtedly provides some means of communication with its visitors. In every place that interaction is possible you have a potential web security vulnerability. Websites often invite visitors to:

- Load a new page containing dynamic content;
- Search for a product or location;
- Fill out a contact form;
- Search the site content;
- Use a shopping cart;
- Create an account.
- Log on to an account.

In each case noted above your website visitor is effectively sending a command to or through your web server—very likely to a database. In each opportunity to communicate, such as a form field, search field or blog, correctly written code will allow only a very narrow range of commands or information types to pass in or out. This is ideal for web security. However, these limits are not automatic. It takes well trained programmers a good deal of time to write code that allows all expected data to pass and disallows[3] all unexpected[4] or potentially harmful data.

And there lies the problem. Code on your site has come from a variety of programmers, some of whom work for third party vendors. Some of that code is old, perhaps very old. Your site may be running software from half a dozen sources, and then your own site designer and your webmaster has each produced more code of their own, or made revisions to the others' code that may have altered or eliminated previously established web security limitations.

Added to that is the software that may have been purchased years ago and which is not in current use. Many servers have accumulated[5] applications that are no longer in use and with which nobody on your current staff is familiar. This code is often not easy to find, is about as valuable as an appendix and has not been used, patched or updated for years—but it may be exactly what a hacker is looking for!

5. Known Web Security Vulnerabilities and Unknown Vulnerabilities

As you know there are a lot of people out there who call themselves hackers. You can also easily guess that they are not all equally skilled. As a matter of fact[6], the vast majority of them are

---

1 punctual *adj.* 严守时刻的，准时的

2 mitigate *v.* 减轻

3 disallow *v.* 不接受

4 unexpected *adj.* 想不到的，意外的，未预料到

5 accumulate *v.* 积聚，堆积

6 as a matter of fact 事实上

simply copycats[1]. They read about a KNOWN technique that was devised by someone else and they use it to break into a site that is interesting to them, often just to see if they can do it. Naturally once they have done that they will take advantage of the site weakness to do malicious harm, plant something or steal something.

A very small number of hackers are actually capable of[2] discovering a new way to overcome web security obstacles[3]. Given the work being done by tens of thousands of programmers worldwide to improve security, it is not easy to discover a brand new method of attack. Hundreds, sometimes thousands of man-hours might be put into developing a new exploit. This is sometimes done by individuals, but just as often is done by teams supported by organized crime. In either case they want to maximize their return on this investment in time and energy and so they will very quietly focus on relatively few, very valuable corporate or governmental assets. Until their new technique is actually discovered, it is considered UNKNOWN.

Countering and attempting to eliminate any return on this hacking investment you have hundreds if not thousands of web security entities[4]. These public and private groups watch for and share information about newly discovered exploits so that an alarm can be raised and defense against unknown exploits can be put in place quickly. The broad announcement of a new exploit makes it a KNOWN exploit.

The outcome of this contest of wills, so to speak, is that exploits become known and widely documented very soon after they are first used and discovered. So at any one time there are thousands (perhaps tens of thousands) of known vulnerabilities and only a very, very few unknown. And those few unknown exploits are very tightly focused onto just a very few highly valuable targets so as to reap the greatest return before discovery. Because once known the best defended sites immediately take action[5] to correct their flaws and erect better defenses.

6. Your Greatest Web Security Risks: Known or Unknown?

Your site is 1,000 times more likely to be attacked with a known exploit than an unknown one. And the reason behind this is simple: There are so many known exploits and the complexity of web servers and websites is so great that the chances are good that one of the known vulnerabilities will be present and allow an attacker access to your site.

The number of sites worldwide is so great and the number of new, as of yet undocumented[6] and thus unknown exploits so small that your chances of being attacked with one is nearly zero—unless you have network assets of truly great value.

If you don't attract the attention of a very dedicated, well financed attack, then your primary concern should be to eliminate your known vulnerabilities so that a quick look would reveal no easy entry using known vulnerabilities.

---

1 copycat *n.* 盲目模仿者

2 be capable of 能够

3 obstacle *n.* 障碍，妨害物

4 entity *n.* 实体

5 take action 采取行动

6 undocumented *adj.* 无正式文件的，无事实证明的

7. Web Security Defense Strategy

There are two roads to accomplish excellent security. On one road, you would assign all of the resources needed to maintain constant alert to new security issues. You would ensure that all patches and updates are done at once, have all of your existing applications reviewed for correct security, ensure that only security knowledgeable programmers do work on your site and have their work checked carefully by security professionals. You would also maintain a tight firewall, antivirus protection and run IPS/IDS.

The other option: Use a web scanning solution to test your existing equipment, applications and website code to see if a KNOWN vulnerability actually exists. While firewalls, antivirus and IPS/IDS are all worthwhile[1], it is simple logic to also lock the front door[2]. It is far more effective to repair a half dozen actual risks than it is to leave them in place and try to build higher and higher walls around them. Network and website vulnerability scanning is the most efficient security investment of all.

If one had to walk just one of these roads, diligent[3] wall building or vulnerability testing, it has been seen that web scanning will actually produce a higher level of web security on a dollar for dollar basis. This is proven by the number of well defended websites which get hacked every month, and the much lower number of properly scanned websites which have been compromised.

8. Web Security Using a Website Security Audit

Your best defense against an attack on your website is to regularly[4] scan a competently set up domain that is running current applications and whose website code was done well.

Website testing, also known as web scanning or auditing, is a hosted service provided by Beyond Security called WSSA—Website Security Audit. This service requires no installation of software or hardware and is done without any interruption of web services.

Beyond Security staff has been accumulating known issues for many years and have compiled what is arguably the world's most complete database of security vulnerabilities. Each kind of exploit has a known combination[5] of website weaknesses that must be present to be accomplished. Thus by examining a server for the open port, available service and/or code that each known exploit requires, it is a simple matter to determine if a server is vulnerable to attack using that method.

In a matter of[6] hours, WSSA can run through its entire database of over ten thousand vulnerabilities and can report on which are present and better yet, confirm the thousands that are not. With that data in hand you and your staff can address your actual web security vulnerabilities and, when handled, know that your site is completely free of known issues regardless of what updates and patches have been done and what condition your code is in or what unused code may reside, hidden, on your site or web server.

---

1 worthwhile *adj.* 值得做的，值得出力的

2 front door 前门

3 diligent *adj.* 勤勉的，用功的，细心而继续不断的

4 regularly *adv.* 有规律地，有规则地

5 combination *n.* 组合，合并

6 a matter of 大约，大概

Then, WSSA can be run on a regular basis so that your site will be tested against new vulnerabilities as they become known and provide you with solid data as to whether action is vital, needed or low priority[1]. You will also be alerted if new code has been added to the site that is insecure[2], a new port has been opened that was unexpected, or a new service has been loaded and started that may present an opportunity to break in.

In complex and large systems it may be that daily web scanning is the ONLY way to ensure that none of the many changes made to site code or on an application may have opened a hole in your carefully established security perimeter!

参考译文

防 火 墙

防火墙是计算机系统或网络的一部分，旨在阻止未经授权的访问，同时允许授权的通信。它是一个设备或设备集合，基于一组规则和其他标准来配置，以允许或拒绝计算机应用（见图 5-1）。

防火墙能够在硬件或软件或两者的组合中实现。防火墙经常用于防止未经授权的互联网用户访问连接到互联网的专用网络，特别是内联网。所有进入或离开内联网的消息都通过防火墙，该防火墙检查每个消息并阻止那些不符合指定安全标准的消息。

防火墙技术有以下几种：

1）数据包过滤：数据包过滤检查通过网络的每个数据包，并根据用户定义的规则接受或拒绝。虽然很难配置，但对用户而言，它相当有效且大多是透明的。它易受 IP 欺骗。

2）应用网关：将安全机制应用于特定应用，如 FTP 和 Telnet 服务器。这非常有效，但可能会降低性能。

3）电路级网关：建立 TCP 或 UDP 连接时应用安全机制。连接完成后，数据包可以在主机之间流动，无须进一步检查。

4）代理服务器：拦截进入和离开网络的所有消息。代理服务器能够有效隐藏真实的网络地址。

1. 功能

防火墙是在计算机上运行的专用设备或软件，用来检查通过它的网络流量，并根据一组规则或条件拒绝或允许其通过。

防火墙通常位于受保护的网络和不受保护的网络之间，并且像保护资产的房门一样，以确保没有秘密信息流出，也没有恶意的信息进入。

防火墙的基本任务是规范不同信任级别的计算机网络之间的一些流量。典型的例子是互联网和内部网络。互联网是一个不可信任的区域，而内部网络是一个具有更高信任度的区域。具有中间信任级别的区域位于互联网和可信内部网络之间，通常被称为“外围网络”或

1 low priority　低优先级

2 insecure　*adj.* 不可靠的，不安全的

非军事区（DMZ）。

网络中的防火墙功能类似于物理防火墙，就像建筑中的防火门。在网络中，防火墙用于防止网络入侵者进入私有网络。在建筑物中，防火门用来挡住和延迟建筑物的火灾蔓延到相邻建筑。

2. 类型

根据通信发生的位置、通信被拦截的位置和正在跟踪的状态，防火墙可以分为以下几种：

（1）网络层防火墙

网络层防火墙（也称数据包过滤器）在 TCP / IP 协议栈的较低层次下运行，它不允许数据包通过防火墙，除非它们与已建立的规则集相匹配。防火墙管理员可以定义规则，或应用默认规则。术语“数据包过滤器”起源于 BSD 操作系统。

网络层防火墙通常分为两个子类：状态防火墙和无状态防火墙。状态防火墙维护活动会话的环境，并使用“状态信息”来加速数据包处理。任何现有的网络连接都可以描述为具有以下几个属性，包括源和目的 IP 地址、UDP 或 TCP 端口以及与生命周期相关的当前阶段（包括会话启动、握手、数据传输或完成连接）。如果数据包与现有连接不匹配，则将根据新连接的规则集进行评估。通过与防火墙的状态表比较，如果数据包与现有连接匹配，则可以通过，而无须进一步处理。

无状态防火墙需要更少的内存，而且对于筛选会话比查找会话需要的时间要少的简单筛选器而言，它们可能更快。无状态防火墙也可能用于过滤不具有会话概念的无状态网络协议。但是，它们不能根据主机之间达到的通信阶段做出更复杂的决策。

现代防火墙可以根据源 IP 地址、源端口、目的 IP 地址或端口、目标服务（如 WWW 或 FTP）等许多数据包属性过滤流量。它们可以基于协议、TTL 值、源的网区、源以及许多其他属性进行过滤。

（2）应用层防火墙

应用层防火墙适用于 TCP/IP 栈（例如，所有浏览器流量或所有 telnet 或 ftp 流量）的应用级，并且可以拦截所有传送到应用或来自应用的数据包。它们阻止其他数据包（通常丢弃它们而不确认发送方）。原则上，应用防火墙可以防止所有不需要的外部流量到达受保护的计算机。

在对所有数据包进行内容正确性检查时，防火墙可以立即限制或阻止网络计算机蠕虫和木马的蔓延。额外的检查标准会延迟数据包转发到其目的地的时间。

（3）代理

代理设备（在专用硬件上运行或作为通用计算机上的软件）可以通过应用的方式响应输入数据包（如连接请求）同时阻止其他数据包，以起到防火墙的作用。

代理使利用外部网络来篡改内部系统更加困难，并且误用一个内部系统不一定会导致来自防火墙以外的可以利用的安全漏洞（只要应用程序代理保持完整并配置正确）。相反，入侵者可能劫持一个公开的系统并将其用作自己的代理，以达到自己的目的；然后代理将伪装成该系统，连接到其他内部机器。虽然内部地址空间的使用增强了安全性，但是骇客仍然可以采用诸如 IP 欺骗等方法来尝试将数据包传递到目标网络。

（4）网络地址转换

防火墙通常具有网络地址转换（NAT）功能，防火墙后面的主机通常具有 RFC 1918 中定义的“私有地址范围”中的地址。防火墙通常具有隐藏受保护主机的真实地址的功能。最初，NAT 功能的开发是为了解决公司或个人的分配和使用 IPv4 可路由地址的限制，也便于减少组织中的每台计算机获得公共地址的数量，并降低相关成本。隐藏受保护设备的地址已经成为越来越重要的网络侦察防御措施。

Unit 6

Text A

Intrusion Detection System

An Intrusion Detection System (IDS) is a device or software application that monitors a network or systems for malicious activity or policy violations. Any detected activity or violation is typically reported either to an administrator or collected centrally using a Security Information and Event Management (SIEM) system. An SIEM system combines outputs from multiple sources, and uses alarm filtering[1] techniques to distinguish malicious activity from false alarms.

There is a wide spectrum of IDS, varying from antivirus software to hierarchical systems that monitor the traffic of an entire backbone network. The most common classifications are Network Intrusion Detection System (NIDS) and Host-based Intrusion Detection System (HIDS)[2]. A system that monitors important operating system files is an example of an HIDS, while a system that analyzes incoming network traffic is an example of an NIDS. It is also possible to classify IDS by detection approach: The most well-known variants are signature-based detection (recognizing bad patterns, such as malware) and anomaly-based detection (detecting deviations from a model of "good" traffic, which often relies on machine learning). Some IDS have the ability to respond to detected intrusions. Systems with response capabilities are typically referred to as an intrusion prevention system.

1. Comparison with Firewalls

Though they both relate to network security, an IDS differs from a firewall in that a firewall looks outwardly for intrusions in order to stop them from happening. Firewalls limit access between networks to prevent intrusion and do not signal an attack from inside the network. An IDS evaluates a suspected intrusion once it has taken place and signals an alarm. An IDS also watches for attacks that originate from within a system. This is traditionally achieved by examining network communications, identifying heuristics and patterns (often known as signatures) of common computer attacks, and taking action to alert operators. A system that terminates connections is called an intrusion prevention system, and is another form of an application layer firewall.

2. Classifications

IDS can be classified by where detection takes place (network or host) and the detection method that is employed.

(1) Analyzed Activity

1) Network Intrusion Detection Systems NIDS. NIDS are placed at a strategic point or points within the network to monitor traffic to and from all devices on the network. It performs an analysis of passing traffic on the entire subnet, and matches the traffic that is passed on the subnets to the library of known attacks. Once an attack is identified, or abnormal behavior is sensed, the alert can be sent to the administrator. An example of an NIDS would be installing it on the subnet where firewalls are located in order to see if someone is trying to break into the firewall. Ideally one would scan all inbound and outbound traffic, however doing so might create a bottleneck that would impair the overall speed of the network. OPNET[3] and NetSim[4] are commonly used tools for simulation network intrusion detection systems. NID Systems are also capable of comparing signatures for similar packets to link and drop harmful detected packets which have a signature matching the records in the NIDS. When we classify the designing of the NIDS according to the system interactivity property, there are two types: on-line and off-line NIDS. On-line NIDS deals with the network in real time. It analyses the Ethernet packets and applies some rules, to decide if it is an attack or not. Off-line NIDS deals with stored data and passes it through some processes to decide if it is an attack or not.

2) Host Intrusion Detection Systems HIDS. HIDS run on individual hosts or devices on the network. An HIDS monitors the inbound and outbound packets from the device only and will alert the user or administrator if suspicious activity is detected. It takes a snapshot of existing system files and matches it to the previous snapshot. If the critical system files were modified or deleted, an alert is sent to the administrator to investigate.

Intrusion detection systems can also be system-specific using custom tools and honeypots[5].

(2) Detection Method

1) Signature-based. Signature-based IDS refers to the detection of attacks by looking for specific patterns, such as byte sequences in network traffic, or known malicious instruction sequences used by malware. This terminology originates from anti-virus software, which refers to these detected patterns as signatures. Although signature-based IDS can easily detect known attacks, it is impossible to detect new attacks, for which no pattern is available.

2) Anomaly-based. Anomaly-based intrusion detection systems were primarily introduced to detect unknown attacks, in part due to the rapid development of malware. The basic approach is to use machine learning to create a model of trustworthy activity, and then compare new behavior against this model. Although this approach enables the detection of previously unknown attacks, it may suffer from false positives: Previously unknown legitimate activity may also be classified as malicious.

New types of what could be called anomaly-based intrusion detection systems are being viewed by Gartner as User and Entity Behavior Analytics (UEBA) (an evolution of the User Behavior Analytics category) and Network Traffic Analysis (NTA). In particular, NTA deals with malicious insiders as well as targeted external attacks that have compromised a user machine or account. Gartner has noted that some organizations have opted for NTA over more traditional IDS.

3. Intrusion Prevention

Some systems may attempt to stop an intrusion attempt but this is neither required nor expected of a monitoring system. Intrusion Detection and Prevention Systems (IDPS) are primarily focused on identifying possible incidents, logging information about them, and reporting attempts. In addition, organizations use IDPS for other purposes, such as identifying problems with security policies, documenting existing threats and deterring individuals from violating security policies. IDPS have become a necessary addition to the security infrastructure of nearly every organization.

IDPS typically record information related to observed events, notify security administrators of important observed events and produce reports. Many IDPS can also respond to a detected threat. They use several response techniques, which involve the IDPS stopping the attack itself, changing the security environment (e.g. reconfiguring a firewall) or changing the attack's content.

Intrusion prevention systems, also known as IDPS, are network security appliances that monitor network or system activities for malicious activity. The main functions of intrusion prevention systems are to identify malicious activity, log information about this activity, report it and attempt to block or stop it.

Intrusion prevention systems are considered extensions of intrusion detection systems because they both monitor network traffic and/or system activities for malicious activity. The main differences are, unlike intrusion detection systems, intrusion prevention systems are placed in-line and are able to actively prevent or block intrusions that are detected. IPS can take such actions as sending an alarm, dropping detected malicious packets, resetting a connection or blocking traffic from the offending IP address. An IPS also can correct Cyclic Redundancy Check (CRC) errors, defragment packet streams, mitigate TCP sequencing issues, and clean up unwanted transport and network layer options.

(1) Classification

Intrusion prevention systems can be classified into four different types:

1) Network-based Intrusion Prevention System (NIPS): Monitors the entire network for suspicious traffic by analyzing protocol activity.

2) Wireless Intrusion Prevention Systems (WIPS): Monitor a wireless network for suspicious traffic by analyzing wireless networking protocols.

3) Network Behavior Analysis (NBA): Examines network traffic to identify threats that generate unusual traffic flows, such as distributed denial of service (DDoS) attacks, certain forms of malware and policy violations.

4) Host-based Intrusion Prevention System (HIPS): An installed software package which monitors a single host for suspicious activity by analyzing events occurring within that host.

(2) Detection Methods

The majority of intrusion prevention systems utilize one of three detection methods: signature-based, statistical anomaly-based, and stateful protocol analysis.

1) Signature-based detection: Signature based IDS monitors packets in the network and

compares with pre-configured and pre-determined attack patterns known as signatures.

2) Statistical anomaly-based detection: An IDS which is anomaly based will monitor network traffic and compare it against an established baseline.

3) Stateful protocol analysis detection: This method identifies deviations of protocol states by comparing observed events with "redetermined profiles of generally accepted definitions of benign activity."

4. Limitations

1) Noise can severely limit an intrusion detection system's effectiveness. Bad packets generated from software bugs, corrupt DNS data, and local packets that escaped can create a significantly high false-alarm rate.

2) It is not uncommon for the number of real attacks to be far below the number of false-alarms. Number of real attacks is often so far below the number of false-alarms that the real attacks are often missed and ignored.

3) Many attacks are geared for specific versions of software that are usually outdated. A constantly changing library of signatures is needed to mitigate threats. Outdated signature databases can leave the IDS vulnerable to newer strategies.

4) For signature-based IDS there will be lag between a new threat discovery and its signature being applied to the IDS. During this lag time the IDS will be unable to identify the threat.

5) It cannot compensate for a weak identification and authentication mechanism or for weaknesses in network protocols. When an attacker gains access due to weak authentication mechanism, IDS cannot prevent the adversary from any malpractice.

6) Encrypted packets are not processed by the intrusion detection software. Therefore, the encrypted packet can allow an intrusion to the network that is undiscovered until more significant network intrusions have occurred.

7) Intrusion detection software provides information based on the network address that is associated with the IP packet that is sent into the network. This is beneficial if the network address contained in the IP packet is accurate. However, the address that is contained in the IP packet could be faked or scrambled.

8) Due to the nature of NIDS systems, and the need for them to analyze protocols as they are captured, NIDS systems can be susceptible to some protocol based attacks that network hosts may be vulnerable. Invalid data and TCP/IP stack attacks may cause an NIDS to crash.

New Words

violation	*n.*	违反，违背，妨碍，侵害
administrator	*n.*	管理员
combine	*v.*	(使)联合，(使)结合
output	*n.*	输出
distinguish	*v.*	区别，辨别
hierarchical	*adj.*	分等级的

backbone	*n.*	中枢，骨干，支柱
anomaly	*n.*	不规则，异常的人或物
respond	*v.*	回答，响应，做出反应
outwardly	*adv.*	表面地，外观上地
evaluate	*v.*	评价，估计，求……的值
suspect	*adj.*	令人怀疑的，不可信的，可疑的
	v.	怀疑，猜想，对……有所觉察
communication	*n.*	通信
pattern	*n.*	式样，模式
	v.	模仿，仿造
signature	*n.*	签名，署名
terminate	*v.*	停止，结束，终止
connection	*n.*	连接，接线
subnet	*n.*	子网络，分支网络
abnormal	*adj.*	反常的，变态的
sense	*v.*	感到，理解，认识
	n.	感觉
inbound	*adj.*	进入的，归航的
	n.	入站
outbound	*adj.*	外出的
bottleneck	*n.*	瓶颈
impair	*v.*	削弱
simulation	*n.*	仿真，模拟
harmful	*adj.*	有害的，伤害的
match	*v.*	使匹配，使相称
Ethernet	*n.*	以太网
host	*n.*	主机
snapshot	*n.*	快照
investigate	*v.*	调查，研究
honeypot	*n.*	蜜罐
byte	*n.*	字节
sequence	*n.*	次序，顺序，序列
trustworthy	*adj.*	可信赖的
suffer	*v.*	遭受，经历，忍受；受痛苦，受损害
incident	*n.*	事件，事变
	adj.	附带的，易于发生的
deter	*v.*	阻止
infrastructure	*n.*	基础设施
observe	*v.*	观察，观测

reconfigure	*v.*	重新配置，改装
reset	*v.*	重新设置
defragment	*n.*	整理碎片
unusual	*adj.*	不平常的，与众不同的，不寻常的
statistical	*adj.*	统计的，统计学的
stateful	*adj.*	有状态的
deviation	*n.*	背离
redetermine	*v.*	重新决定，再决定
corrupt	*adj.*	被破坏的
uncommon	*adj.*	罕有的，难得的
ignore	*v.*	不理睬，忽视
gear	*v.*	调整，(使)适合
outdated	*adj.*	过时的，不流行的
constantly	*adv.*	不变地，经常地
lag	*n.*	落后
	v.	滞后，落后于
compensate	*v.*	偿还，补偿
adversary	*n.*	敌手，对手
malpractice	*n.*	弊端，失职
undiscovered	*adj.*	未被发现的，隐藏的
accurate	*adj.*	正确的，精确的
scramble	*n.*	混乱
	v.	杂乱蔓延，拼凑，搅乱，使混杂
invalid	*adj.*	无效的
crash	*v.*	崩溃

Phrases

alarm filtering	报警过滤
distinguish...from...	把……和……区分开来
a wide spectrum of	各种各样的，许多的，多种的
response capability	响应能力，应对能力，反应能力
false alarm	假警报，误报警
originate from	发源于，起源于
match to	与……匹配
byte sequence	字节序列
instruction sequence	指令序列，控制序列
false positive	假阳性
legitimate activity	合法的活动
in particular	尤其，特别

look for	寻找
deal with	处理，涉及
opt for	选择
focus on	致力于，使聚焦于，对（某事）予以注意
deter sb. from	阻止某人做某事
network security appliance	网络安全设备，网络安全设施
packet stream	数据包流
clean up	清理，整理
compensate for	弥补
authentication mechanism	认证机制，验证机制

Abbreviations

IDS (Intrusion Detection System)	入侵检测系统
SIEM (Security Information and Event Management)	安全信息和事件管理
NIDS (Network Intrusion Detection System)	网络入侵检测系统
HIDS (Host-based Intrusion Detection System)	基于主机的入侵检测系统
UEBA (User and Entity Behavior Analytics)	用户和实体行为分析
NTA (Network Traffic Analysis)	网络流量分析
IDPS (Intrusion Detection and Prevention System)	入侵检测和预防系统
CRC (Cyclic Redundancy Check)	循环冗余码校验
NIPS (Network-based Intrusion Prevention System)	基于网络的入侵预防系统
NBA (Network Behavior Analysis)	网络行为分析
DNS (Domain Name Server)	域名服务器

Notes

[1] Alarm filtering, in the context of IT network management, is the method by which an alarm system reports the origin of a system failure, rather than a list of systems failed.

[2] A Host-based Intrusion Detection System (HIDS) is an intrusion detection system that monitors and analyzes the internals of a computing system as well as (in some cases) the network packets on its network interfaces [just like a Network-based Intrusion Detection System (NIDS) would do]. This was the first type of intrusion detection software to have been designed, with the original target system being the mainframe computer where outside interaction was infrequent.

[3] OPNET Technologies, Inc. was a software business that provided performance management for computer networks and applications.

[4] NetSim is a network simulation and network emulation tool used for network design & planning, defense applications and network R & D. Various technologies such as Cognitive Radio, Wireless Sensor Networks, Wireless LAN, Wi Max, MANETs, Wireless Sensor Networks, LTE, etc. are covered in NetSim.

[5] In computer terminology, a honeypot is a computer security mechanism set to detect, deflect, or, in some manner, counteract attempts at unauthorized use of information systems. Generally, a honeypot consists of data (for example, in a network site) that appears to be a legitimate part of the site, but is actually isolated and monitored, and that seems to contain information or a resource of value to attackers, who are then blocked. This is similar to the police baiting a criminal.

Exercises

[Ex. 1] Answer the following questions according to the text.

1. What is an Intrusion Detection System (IDS)?
2. What are the most common classifications of IDS? What are the examples of each of them?
3. What is the difference between an IDS and a firewall?
4. What do NIDS do?
5. Where do host intrusion detection systems (HIDS) run? What does an HIDS do?
6. What does signature-based IDS refer to? Where does this terminology originate from?
7. What are Intrusion Detection and Prevention Systems (IDPS) doing? What other purposes do organizations use IDPS for?
8. What are the main functions of intrusion prevention systems?
9. How many different types can intrusion prevention systems be classified into? What are they?
10. What are the three detection methods mentioned in the text?

[Ex. 2] Translate the following terms or phrases from English into Chinese and vice versa.

1. alarm filtering	1. ______
2. network security appliance	2. ______
3. authentication mechanism	3. ______
4. false alarm	4. ______
5. distinguish...from...	5. ______
6. 反常的，变态的	6. ______
7. 瓶颈	7. ______
8. 连接，接线	8. ______
9. 蜜罐	9. ______
10.重新配置，改装	10. ______

[Ex. 3] Translate the following passage into Chinese.

Vulnerability Scanner

A vulnerability scanner is a computer program designed to assess computers, computer systems, networks or applications for weaknesses.

1. What is a vulnerability scan?

The bad guys are constantly scanning for vulnerabilities in systems out there, to find out who is already open to attack. The idea of a vulnerability scan is that you find the weaknesses and fix them before they get to you. Whether it's in your network, infrastructure, or web application, a vulnerability is normally a flaw in your code, which attackers will use to break into your system before proceeding to totally rinse you.

2. Why do you need one?

Aside from letting you know where you need to reinforce your system, thereby protecting you and your customers, it can help with becoming PCI-compliant.

3. How does it work?

The scan digs through all your ports, processes, firewall policies and software updates and can also be performed for individual applications.

4. What is it looking for?

All sorts of nasties, but it's basically stopping attacks by finding new vulnerabilities, as well as giving advice on your security, and any patches available etc. It also lets you know the types of threats and potential threats, and can even provide training materials for your business.

[Ex. 4] Fill in the blanks with the words given below.

predefined	limited	implementing	attacking	discovering
scanners	variety	cryptographic	configured	weaknesses

Web Application Security Scanner

A web application security scanner is program which communicates with a web application through the web front-end in order to identify potential security vulnerabilities in the web application and architectural weaknesses. It performs a black-box test. Unlike source code scanners, web application ___1___ don't have access to the source code and therefore detect vulnerabilities by actually performing attacks.

1. Overview

A web application security scanner can facilitate the automated review of a web application with the expressed purpose of ___2___ security vulnerabilities, and are required to comply with various regulatory requirements. Web application scanners can look for a wide ___3___ of vulnerabilities, including:

- Input/Output validation: (Cross-site scripting, SQL Injection, etc.);
- Specific application problems;
- Server configuration mistakes/errors/version.

2. Strengths and Weaknesses

Like every testing tools, the web application security scanner is not a perfect tool, it has strengths and ___4___.

(1) Weaknesses and Limitations

- Because the tool is ___5___ a dynamic testing method, it cannot cover 100% of the source code of the application and then, the application itself. The penetration tester should look at the coverage of the web application or of its attack surface to know if the tool was ___6___ correctly or was able to understand the web application.
- It is really hard for a tool to find logical flaws such as the use of weak ___7___ functions, information leakage, etc.
- Even for technical flaws, if the web application doesn't give enough clue, the tool cannot

catch them.

- The tool cannot implement all variants of attacks for a given vulnerability. So the tools generally have a ___8___ list of attacks and does not generate the attack payloads depending on the tested web application.
- The tools are usually ___9___ in the understanding of the application behavioral with dynamic content such as JavaScript, Flash, etc.

(2) Strengths

- The tool can detect vulnerabilities of the finalized release candidate before shipping.
- It simulates a malicious user by ___10___ and probing, and seeing what results are not part of the expected result set.
- As a dynamic testing tool, it is not language dependent. A web application scanner is able to scan Java/JSP, PHP or any other engine driven web application.

Text B

Port Scanner

Port scanner is an application designed to probe a server or host for open ports. This is often used by administrators to verify security policies of their networks and by attackers to identify services running on a host and exploit vulnerabilities.

A port scan or portscan is a process that sends client requests to a range of server port addresses on a host, with the goal of finding an active port; this is not a nefarious process in and of itself. The major use of a port scan is not to attack but rather simply to probe to determine services available on a remote machine.

To portsweep is to scan multiple hosts for a specific listening port. The latter is typically used to search for a specific service. For example, an SQL-based computer worm may portsweep looking for hosts listening on TCP port 1433.

1. TCP/IP Basic Knowledge

The design and operation of the Internet is based on the Internet Protocol Suite, commonly also called TCP/IP. In this system, hosts and host services are referenced using two components: an address and a port number. There are 65536 distinct and usable port numbers. Most services use a limited range of numbers.

Some port scanners scan only the most common port numbers, or ports most commonly associated with vulnerable services, on a given host.

The result of a scan on a port is usually generalized into one of three categories:

1) Open or Accepted: The host sent a reply indicating that a service is listening on the port.

2) Closed or Denied or Not Listening: The host sent a reply indicating that connections will be denied to the port.

3) Filtered, Dropped or Blocked: There was no reply from the host.

Open ports present two vulnerabilities of which administrators must be wary:

1) Security and stability concerns associated with the program responsible for delivering the service—Open ports.

2) Security and stability concerns associated with the operating system that is running on the host—Open or Closed ports.

Filtered ports do not tend to present vulnerabilities.

2. Assumptions

All forms of port scanning rely on the assumption that the targeted host is compliant with RFC 793—Transmission Control Protocol. Although this is the case most of the time, there is still a chance a host might send back strange packets or even generate false positives when the TCP/IP stack of the host is non-RFC[1]-compliant or has been altered. This is especially true for less common scan techniques that are OS-dependent. The TCP/IP stack fingerprinting method also relies on these types of different network responses from a specific stimulus to guess the type of the operating system the host is running.

3. Types

(1) TCP Scanning

The simplest port scanners use the operating system's network functions and are generally the next option to go to when SYN is not a feasible option. Nmap calls this mode connect scan, named after the Unix connect() system call. If a port is open, the operating system completes the TCP three-way handshake, and the port scanner immediately closes the connection to avoid performing a denial-of-service attack. Otherwise an error code is returned. This scan mode has the advantage that the user does not require special privileges. However, using the OS network functions prevents low-level control, so this scan type is less common. This method is "noisy", particularly if it is a "portsweep": The services can log the sender IP address and intrusion detection systems can raise an alarm.

(2) SYN Scanning

SYN scan is another form of TCP scanning. Rather than use the operating system's network functions, the port scanner generates raw IP packets itself, and monitors for responses. This scan type is also known as "half-open scanning", because it never actually opens a full TCP connection. The port scanner generates an SYN packet. If the target port is open, it will respond with an SYN-ACK packet. The scanner host responds with an RST packet, closing the connection before the handshake is completed. If the port is closed but unfiltered, the target will instantly respond with an RST packet.

The use of raw networking has several advantages, giving the scanner full control of the packets sent and the timeout for responses, and allowing detailed reporting of the responses. There is debate over which scan is less intrusive on the target host. SYN scan has the advantage that the individual services never actually receive a connection. However, the RST during the handshake can cause problems for some network stacks, in particular simple devices like printers. There are no conclusive arguments either way.

(3) UDP[2] Scanning

UDP scanning is also possible, although there are technical challenges. UDP is a connectionless protocol so there is no equivalent to a TCP SYN packet. However, if a UDP packet is sent to a port that is not open, the system will respond with an ICMP[3] port unreachable message. Most UDP port scanners use this scanning method, and use the absence of a response to infer that a port is open. However, if a port is blocked by a firewall, this method will falsely report that the port is open. If the port unreachable message is blocked, all ports will appear open. This method is also affected by ICMP rate limiting.

An alternative approach is to send application-specific UDP packets, hoping to generate an application layer response. For example, sending a DNS query to port 53 will result in a response, if a DNS server is present. This method is much more reliable at identifying open ports. However, it is limited to scanning ports for which an application specific probe packet is available. Some tools (e.g., nmap) generally have probes for less than 20 UDP services, while some commercial tools (e.g., nessus) have as many as 70. In some cases, a service may be listening on the port, but configured not to respond to the particular probe packet.

(4) ACK Scanning

ACK scanning is one of the more unusual scan types, as it does not exactly determine whether the port is open or closed, but whether the port is filtered or unfiltered. This is especially good when attempting to probe for the existence of a firewall and its rulesets. Simple packet filtering will allow established connections (packets with the ACK bit set), whereas a more sophisticated stateful firewall might not.

(5) Window Scanning

Rarely used because of its outdated nature, window scanning is fairly untrustworthy in determining whether a port is opened or closed. It generates the same packet as an ACK scan, but checks whether the window field of the packet has been modified. When the packet reaches its destination, a design flaw attempts to create a window size for the packet if the port is open, flagging the window field of the packet with 1's before it returns to the sender. Using this scanning technique with systems that no longer support this implementation returns 0's for the window field, labeling open ports as closed.

(6) FIN Scanning

Since SYN scans are not surreptitious enough, firewalls are, in general, scanning for and blocking packets in the form of SYN packets. FIN packetscan bypass firewalls without modification. Closed ports reply to an FIN packet with the appropriate RST packet, whereas open ports ignore the packet on hand. This is typical behavior due to the nature of TCP, and is in some ways an inescapable downfall.

(7) Other Scan Types

Some more unusual scan types exist. These have various limitations and are not widely used. Nmap supports most of these.

1) X-mas and Null Scan—are similar to FIN scanning, but:

① X-mas sends packets with FIN, URG and PUSH flags turned on like a Christmas tree.

② Null sends a packet with no TCP flags set.

2) Protocol scan—determines what IP level protocols (TCP, UDP, GRE, etc.) are enabled.

3) Proxy scan—A proxy (SOCKS or HTTP) is used to perform the scan. The target will see the proxy's IP address as the source. This can also be done using some FTP servers.

4) Idle scan—Another method of scanning without revealing one's IP address, taking advantage of the predictable IP ID flaw.

5) CatSCAN—checks ports for erroneous packets.

6) ICMP scan—determines if a host responds to ICMP requests, such as echo (ping), netmask, etc.

4. Port Filtering by ISPs

Many Internet service providers restrict their customers' ability to perform port scans to destinations outside of their home networks. This is usually covered in the terms of service or acceptable use policy to which the customer must agree. Some ISPs implement packet filters or transparent proxies that prevent outgoing service requests to certain ports. For example, if an ISP provides a transparent HTTP proxy on port 80, port scans of any address will appear to have port 80 open, regardless of the target host's actual configuration.

5. Ethics

The information gathered by a port scan has many legitimate uses including network inventory and the verification of the security of a network. Port scanning can, however, also be used to compromise security. Many exploits rely upon port scans to find open ports and send specific data patterns in an attempt to trigger a condition known as a buffer overflow. Such behavior can compromise the security of a network and the computers therein, resulting in the loss or exposure of sensitive information and the ability to do work.

The threat level caused by a port scan can vary greatly according to the method used to scan, the kind of port scanned, its number, the value of the targeted host and the administrator who monitors the host. But a port scan is often viewed as a first step for an attack, and is therefore taken seriously because it can disclose much sensitive information about the host. Despite this, the probability of a port scan alone followed by a real attack is small. The probability of an attack is much higher when the port scan is associated with a vulnerability scan.

New Words

scanner	*n.*	扫描器，扫描仪
probe	*n.*	探针，探测器
	v.	探查，查明
nefarious	*adj.*	邪恶的，穷凶极恶的
remote	*adj.*	远程的，遥远的
generalized	*adj.*	广泛的，普遍的

reply	*n.*	答复
	v.	答复，回答
indicate	*v.*	指示，表明，表示
drop	*v.*	放弃，停止
stability	*n.*	稳定性
compliant	*adj.*	兼容的，适应的
strange	*adj.*	陌生的，生疏的，前所未知的，奇怪的
stimulus	*n.*	刺激
feasible	*adj.*	可行的，切实可行的
call	*v.*	调用
privilege	*n.*	特权，特别待遇
	v.	给予……特权
unfiltered	*adj.*	未滤过的
timeout	*n.*	超时
intrusive	*adj.*	打扰的，插入的
unreachable	*adj.*	不能达到的，不能得到的
message	*n.*	消息
	v.	通知，带信息
infer	*v.*	推断
ruleset	*n.*	规则集
untrustworthy	*adj.*	不能信赖的，靠不住的
flag	*n. & v.*	标记
field	*n.*	域
label	*n.*	标签，标志
	v.	贴标签于，分类，标注
surreptitious	*adj.*	暗中的，秘密的
inescapable	*adj.*	逃不掉的，不可避免的
downfall	*n.*	衰败，垮台
restrict	*v.*	限制，约束，限定
proxy	*n.*	代理，代理人
trigger	*v.*	引发，引起，触发
exposure	*n.*	暴露，揭露，曝光
probability	*n.*	可能性，或然性，概率

Phrases

open port	开放端口
port scan	端口扫描
active port	活动端口
Internet Protocol Suite	因特网协议族

port number	端口号
send back	退还
system call	系统调用
three-way handshake	三次握手
connectionless protocol	无连接协议
be equivalent to...	相当于……，等同于，与……等效
buffer overflow	缓存溢出，缓冲区溢出

Abbreviations

RFC (Request For Comments)	请求注解，Internet 标准(草案)
ICMP (Internet Control Messages Protocol)	网间控制报文协议
FTP (File Transfer Protocol)	文件传输协议
ISP (Internet Service Provider)	因特网服务提供商

Notes

[1] A Request for Comments (RFC) is a type of publication from the Internet Engineering Task Force (IETF) and the Internet Society (ISOC), the principal technical development and standards-setting bodies for the Internet.

[2] User Datagram Protocol or UDP is part of the Internet Protocol suite. By using UDP, programs running on different computers on a network can send short messages known as Datagrams to one another. UDP can be used in networks where TCP is traditionally implemented, but unlike TCP, it does not guarantee reliability or correct data sequencing. Datagrams may go missing without notice or arrive in a different order from how they were sent.

[3] The Internet Control Message Protocol (ICMP) is a supporting protocol in the Internet protocol suite. It is used by network devices, including routers, to send error messages and operational information indicating, for example, that a requested service is not available or that a host or router could not be reached. ICMP differs from transport protocols such as TCP and UDP in that it is not typically used to exchange data between systems, nor is it regularly employed by end-user network applications (with the exception of some diagnostic tools like ping and traceroute).

Exercises

[Ex. 5] Answer the following questions according to the text.

1. What is port scanner? What is it often used to do by administrators?
2. What is a port scan?
3. What are the three categories of the result of a scan on a port?
4. What are the two vulnerabilities open ports present?
5. What do the simplest port scanners use?
6. What is SYN scan? What is it also known as and why?
7. What will the system do if a UDP packet is sent to a port that is not open?

8. Why is ACK scanning one of the more unusual scan types?

9. What is the difference between closed ports and open ports?

10. What is a port scan often viewed as? Why is it taken seriously?

Reading Material

Database Security

Database[1] security concerns[2] the use of a broad range of information security controls to protect databases (potentially including the data, the database applications or stored functions, the database systems, the database servers and the associated network links) against compromises of their confidentiality, integrity[3] and availability[4]. It involves various types or categories of controls, such as technical, procedural/administrative and physical. Database security is a specialist topic within the broader realms of computer security, information security and risk management[5].

1. Security Risks to Database Systems

Some of the security risks to database systems are as follows.

- Unauthorized or unintended[6] activity or misuse by authorized database users, database administrators, or network/systems managers, or by unauthorized users or hackers (e.g. inappropriate[7] access to sensitive data, metadata[8] or functions within databases, or inappropriate changes to the database programs, structures[9] or security configurations).
- Malware infections causing incidents such as unauthorized access, leakage[10] or disclosure of personal or proprietary data, deletion[11] of or damage to the data or programs, interruption or denial of authorized access to the database, attacks on other systems and the unanticipated failure of database services.
- Overloads[12], performance constraints and capacity issues resulting in the inability[13] of authorized users to use databases as intended.
- Physical damage[14] to database servers caused by computer room fires or floods, overheating,

---

1 database　*n.*　数据库

2 concern　*v.*　涉及，关系到

3 integrity　*n.*　完整性

4 availability　*n.*　可用性，有效性，实用性

5 risk management　风险管理

6 unintended　*adj.*　非故意的，无意识的

7 inappropriate　*adj.*　不适当的，不相称的

8 metadata　*n.*　元数据

9 structure　*n.*　结构，构造

10 leakage　*n.*　漏，泄漏，渗漏

11 deletion　*n.*　删除

12 overload　*n.*　超载，负荷过多

13 inability　*n.*　无能，无力

14 physical damage　物理损坏，有形损坏

lightning[1], accidental liquid spills, static discharge[2], electronic breakdowns/equipment failures and obsolescence[3].

- Design flaws and programming bugs in databases and the associated programs and systems, creating various security vulnerabilities (e.g. unauthorized privilege escalation[4]), data loss/corruption, performance degradation[5] etc.
- Data corruption and/or loss caused by the entry of invalid data[6] or commands, mistakes in database or system administration processes, sabotage[7]/criminal damage etc.

Ross J. Anderson has often said that by their nature large databases will never be free of abuse by breaches of security; if a large system is designed for ease of access it becomes insecure; if made watertight it becomes impossible to use. This is sometimes known as Anderson's Rule.

Many layers and types of information security control are appropriate to databases, including:

- Access control;
- Auditing;
- Authentication;
- Encryption;
- Integrity controls;
- Backups[8];
- Application security;
- Database Security applying Statistical Method.

2. Vulnerability Assessments to Manage Risk and Compliance

One technique for evaluating database security involves performing vulnerability assessments or penetration[9] tests against the database. Testers attempt to find security vulnerabilities that could be used to defeat or bypass security controls, break into the database, compromise the system etc. Database administrators or information security administrators may, for example, use automated vulnerability scans to search out misconfiguration of controls within the layers mentioned above along with known vulnerabilities within the database software. The results of such scans are used to harden the database (improve security) and close off the specific vulnerabilities identified, but other vulnerabilities often remain unrecognized[10] and unaddressed.

In database environments where security is critical, continual[11] monitoring for compliance with

---

1 lightning *n.* 闪电
2 static discharge 静电放电
3 obsolescence *n.* 荒废，退化
4 escalation *n.* 扩大，增加
5 degradation *n.* 退化
6 invalid data 无效数据
7 sabotage *n.* 破坏活动
8 backup *n.* 做备份
9 penetration *n.* 穿过，渗透，突破
10 unrecognized *adj.* 不能辨认的
11 continual *adj.* 连续的，频繁的，持续不断的

standards improves security. Security compliance requires, among other procedures, patch management and the review and management of permissions (especially public) granted to objects[1] within the database. Database objects may include table[2] or other objects listed in the Table link. The permissions[3] granted for SQL language commands on objects are considered in this process. Compliance monitoring is similar to vulnerability assessment, except that the results of vulnerability assessments generally drive the security standards that lead to the continuous monitoring program. Essentially, vulnerability assessment is a preliminary[4] procedure to determine risk where a compliance program is the process of on-going risk assessment.

The compliance program should take into consideration[5] any dependencies at the application software level as changes at the database level may have effects on the application software or the application server.

3. Abstraction[6]

Application level authentication and authorization[7] mechanisms may be effective means of providing abstraction from the database layer. The primary benefit of abstraction is that of a single sign-on[8] capability across multiple databases and platforms[9]. A single sign-on system stores the database user's credentials and authenticates to the database on behalf of[10] the user.

4. Database Activity Monitoring (DAM)

Another security layer of a more sophisticated nature includes real-time database activity monitoring, either by analyzing protocol traffic (SQL) over the network, or by observing local database activity on each server using software agents, or both. Use of agents or native logging is required to capture activities executed on the database server, which typically include the activities of the database administrator. Agents allow this information to be captured in a fashion that can not be disabled by the database administrator, who has the ability to disable or modify native audit logs.

Analysis can be performed to identify known exploits or policy breaches, or baselines[11] which can be captured over time to build a normal pattern used for detection of anomalous activity that could be indicative of intrusion. These systems can provide a comprehensive[12] database audit trail in addition to the intrusion detection mechanisms, and some systems can also provide

---

1 object *n.* 对象

2 table *n.* 表

3 permission *n.* 许可，允许

4 preliminary *adj.* 预备的，初步的

5 take into consideration 考虑到

6 abstraction *n.* 提取

7 authorization *n.* 授权，认可

8 Single sign-on (SSO, 单点登录) is a property of access control of multiple related, yet independent, software systems. With this property, a user logs in with a single ID and password to gain access to a connected system or systems without using different usernames or passwords, or in some configurations seamlessly sign on at each system.

9 platform *n.* 平台

10 on behalf of 代表

11 baseline *n.* 基线

12 comprehensive *adj.* 全面的，广泛的

protection by terminating user sessions[1] and/or quarantining[2] users demonstrating suspicious behavior. Some systems are designed to support Separation[3] of Duties (SOD), which is a typical requirement of auditors. SOD requires that the database administrators who are typically monitored as part of the DAM, not be able to disable or alter the DAM functionality. This requires the DAM audit trail to be securely stored in a separate[4] system not administered by the database administration group.

5. Native Audit

In addition to using external tools for monitoring or auditing, native database audit capabilities are also available for many database platforms. The native audit trails are extracted on a regular basis and transferred to a designated security system where the database administrators do/should not have access. This ensures a certain level of segregation of duties that may provide evidence the native audit trails were not modified by authenticated administrators, and should be conducted by a security-oriented[5] senior DBA group with read rights into production. Turning on native impacts the performance of the server.

6. Process and Procedures

A good database security program includes the regular review of privileges granted to user accounts and accounts used by automated processes. For individual accounts a two-factor authentication[6] system improves security but adds complexity and cost. Accounts used by automated processes require appropriate controls around password storage such as sufficient encryption and access controls to reduce the risk of compromise.

In conjunction with[7] a sound database security program, an appropriate Disaster Recovery (DR)[8] program can ensure that service is not interrupted during a security incident[9], or any incident incident that results in an outage of the primary database environment. An example is that of replication[10] for the primary databases to sites located in different geographical regions.

---

1 session *n.* 会话

2 quarantine *n. & v.* 隔离，封锁

3 separation *n.* 分离，分开

4 separate *adj.* 分开的，分离的

5 oriented *adj.* 导向的，面向……的

6 Two-Factor Authentication (also known as 2FA) is a method of confirming a user's claimed identity by utilizing a combination of two different components. Two-factor authentication is a type of multi-factor authentication.

A good example from everyday life is the withdrawing of money from a cash machine; only the correct combination of a bank card (something that the user possesses) and a PIN (personal identification number, something that the user knows) allows the transaction to be carried out.

7 in conjunction with 与……协力

8 Disaster Recovery (DR) involves a set of policies, procedures and tools to enable the recovery or continuation of vital technology infrastructure and systems following a natural or human-induced disaster. Disaster recovery focuses on the IT or technology systems supporting critical business functions, as opposed to business continuity, which involves keeping all essential aspects of a business functioning despite significant disruptive events. Disaster recovery is therefore a subset of business continuity.

9 incident *n.* 事件，事变

10 replication *n.* 复制

After an incident occurs, database forensics[1] can be employed to determine the scope of the breach, and to identify appropriate changes to systems and processes.

参 考 译 文

入侵检测系统

入侵检测系统（IDS）是一个设备或软件应用程序，用来监控网络或系统的恶意或违反政策的活动。使用安全信息和事件管理（SIEM）系统检测到的任何活动或违规行为通常报告给管理员或集中收集。SIEM 系统组合来自多个源的输出，并使用报警过滤技术来区分恶意活动和误报。

IDS 有许多种类，涉及从防病毒软件到监控整个骨干网络流量的分层系统。最常见的有网络入侵检测系统（NIDS）和基于主机的入侵检测系统（HIDS）。HIDS 是监控重要操作系统文件的系统的一个范例，而 NIDS 是分析进入网络流量的系统的一个范例。还可以根据检测方法对 IDS 进行分类：最知名的是基于签名的检测（识别错误的模式，如恶意软件）和基于异常的检测（检测与“良好”流量模型的偏差，这通常要依靠机器学习）。一些 IDS 能够响应检测到的入侵。具有响应能力的系统通常被称为入侵防御系统。

1. 与防火墙进行比较

虽然它们都与网络安全性有关，但 IDS 与防火墙的不同之处在于，防火墙寻找来自外部的入侵，以阻止其发生。防火墙限制网络之间的访问，以防止入侵并且不对网络内的攻击进行报警。IDS 评估怀疑的入侵，一旦有此类入侵就发出警报。IDS 还会监控来自系统内的攻击。通常，通过检查网络通信，用启发式和模式识别（通常称为签名）就会发现普通计算机攻击，并向操作员报警。终止连接的系统称为入侵防御系统，是应用层防火墙的另一种形式。

2. 分类

可以通过检测的位置（网络或主机）和所采用的检测方法对 IDS 进行分类。

（1）分析活动

1）网络入侵检测系统（NIDS）。NIDS 放置在网络中的战略点或网内各点，以监控网络上所有设备的进出流量。它对整个子网上的流量进行分析，并将子网上传递的流量与已知攻击库进行比对。一旦识别出攻击或感觉到异常行为，就可以向管理员报警。可以将 NIDS 安装在防火墙所在的子网上，以查看某人是否试图进入防火墙。在理想情况下，它将扫描所有入站和出站流量，但这样做可能会造成一个瓶颈，从而影响网络的整体速度。 OPNET 和 NetSim 是模拟网络入侵检测系统的常用工具。NID 系统还能够比较链接的相似数据包签名，并丢弃已经检测到的签名与 NIDS 中记录匹配的有害数据包。当我们根据系统交互性属性对 NIDS 的设计进行分类时，可以分为两种类型：在线 NIDS 和离线 NIDS。在线 NIDS 实时处理网络用于分析以太网数据包并应用一些规则，以确定是否属于攻击。离线 NIDS 处理

1 Database forensics is a branch of digital forensic science relating to the forensic study of databases and their related metadata. The discipline is similar to computer forensics, following the normal forensic process and applying investigative techniques to database contents and metadata.

存储的数据，并通过一些过程来确定是否属于一个攻击。

2）主机入侵检测系统（HIDS）。HIDS 在网络上的各个主机或设备上运行。HIDS 仅监控来自设备的入站和出站数据包，并且如果检测到可疑活动，则会提醒用户或管理员。它需要现有系统文件的快照，并将其与先前的快照进行匹配。如果关键系统文件被修改或删除，则会向管理员报警以进行调查。

入侵检测系统也可以使用用户工具和蜜罐的定制系统。

（2）检测方法

1）基于签名。基于签名的 IDS 是指通过查找特定模式（如网络流量中的字节序列）或恶意软件使用的已知恶意指令序列来检测攻击。这个术语源于反病毒软件，它将这些检测到的模式称为签名。虽然基于签名的 IDS 可以轻松地检测到已知的攻击，但是对于新的攻击，因为没有可用的模式，所以无法检测。

2）基于异常。基于异常的入侵检测系统主要用于检测未知攻击，部分原因是恶意软件的快速发展。基本的方法是使用机器学习创建可信赖的活动模型，然后把新的行为与这个模型比较。虽然这种方法可以检测以前未知的攻击，但可能会遭受误报（假阳性）：以前未知的合法活动也可能被归类为恶意攻击。

Gartner 将新类型的异常入侵检测系统视为用户和实体行为分析（UEBA）（用户行为分析类别的发展）和网络流量分析（NTA）。尤其是，NTA 处理恶意内部人员以及已经破坏用户计算机或账户的有针对性的外部攻击。Gartner 指出，一些组织已经选择了 NTA 而不是更传统的 IDS。

3. 入侵防范

某些系统可能会尝试停止入侵企图，但这不是监控系统所必需的，也不是期望的。入侵检测和预防系统（IDPS）主要侧重于识别可能的事件，记录有关它们的信息和报告企图。此外，组织还可以使用 IDPS 实现其他目的，例如确定安全策略的问题，记录现有威胁并阻止个人违反安全策略。IDPS 已经成为几乎每个组织的安全基础设施的必要补充。

IDPS 通常记录与观察事件相关的信息、通知安全管理员观察到的重要事件并生成报告。许多 IDPS 也能够对检测到的威胁做出响应。它们使用几种响应技术，包括 IDPS 停止攻击本身、改变安全环境（如重新配置防火墙）或改变攻击的内容。

入侵防御系统（IPS）也被称为入侵检测和预防系统（IDPS），是监控恶意网络活动或系统活动的网络安全应用设施。入侵防御系统的主要功能是识别恶意活动，记录有关此活动的信息，报告并尝试阻止或停止。

入侵防御系统被认为是入侵检测系统的扩展，因为它们都监控恶意活动的网络流量和/或系统活动。与入侵检测系统主要不同之处在于，入侵防御系统是在线的并且能够主动地防止或阻止被检测到的入侵。IPS 可以采取的行动有：发送警报、丢弃检测到的恶意数据包、重置连接或阻止来自违规 IP 地址的流量。IPS 还可以校正循环冗余校验（CRC）错误、对数据包流进行碎片整理、减轻 TCP 排序问题并清理不需要的传输和网络层选项。

（1）分类

入侵防御系统可分为四种类型：

1）基于网络的入侵防御系统（NIPS）：通过分析协议活动来监控整个网络的可疑流量。

2）无线入侵防御系统（WIPS）：通过分析无线网络协议来监控无线网络的可疑流量。

3）网络行为分析（NBA）：检查网络流量，以识别产生异常流量的威胁，例如分布式拒绝服务（DDoS）攻击、某些形式的恶意软件和违反政策的行为。

4）基于主机的入侵防御系统（HIPS）：安装的软件包，通过分析在该主机内发生的事件来监控单个主机的可疑活动。

（2）检测方法

大多数入侵防御系统使用以下三种检测方法之一：

1）基于签名的检测：基于签名的 IDS 监控网络中的数据包，并将其与预先配置的和预定义的称为签名的攻击模式进行比对。

2）基于统计异常的检测：基于异常的 IDS 监控网络流量并将其与已建立的基线进行比较。

3）状态协议分析检测：该方法识别协议状态的偏差，方式是将观察到的事件与“重新定义的普遍接受的良性活动进行比较。

4. 限制

1）噪声可严重限制入侵检测系统的有效性。从软件缺陷产生的错误数据包、损坏的 DNS 数据以及逃逸的本地数据包都可能会产生很高的误报率。

2）真实攻击的数量远远低于虚假警报数量，这并不罕见。真实攻击的数量通常远远低于虚假警报数量，以致真正的攻击往往被错过和忽视。

3）许多攻击通常适用于过时的特定版本的软件。需要不断改变签名库来减轻威胁。过时的签名数据库可能使 IDS 容易受到较新策略的攻击。

4）对于基于签名的 IDS，应用于 IDS 的签名滞后于新发现的威胁。在这个滞后期间，IDS 将无法识别该威胁。

5）它无法补偿弱识别和认证机制或网络协议的弱点。当攻击者由于身份验证机制较弱而获得访问权限时，IDS 无法阻止其进行任何不法行为。

6）入侵检测软件不会对加密的数据包进行处理。因此，加密的数据包可以侵入网络而不被发现，直到发生更严重的网络入侵。

7）入侵检测软件基于网络地址提供信息，这些地址与发送到该网络的 IP 包相关联。如果包含在 IP 包中的网络地址是准确的，这没问题。然而，IP 包中包含的地址可能被伪造或搅乱。

8）由于 NIDS 系统的性质并且其在捕获协议时需要分析协议，NIDS 系统可能会受到一些基于协议的攻击，网络主机也易受其攻击。无效的数据和 TCP/IP 栈攻击可能会导致 NIDS 崩溃。

Unit 7

Text A

Encryption and Decryption

Encryption is the process of transforming information so it is unintelligible to anyone but the intended recipient. Decryption is the process of transforming encrypted information so that it is intelligible again. A cryptographic algorithm, also called a cipher, is a mathematical function used for encryption or decryption. In most cases, two related functions are employed, one for encryption and the other for decryption.

With most modern cryptography, the ability to keep encrypted information secret is based not on the cryptographic algorithm, which is widely known, but on a number called a key that must be used with the algorithm to produce an encrypted result or to decrypt previously encrypted information. Decryption with the correct key is simple. Decryption without the correct key is very difficult, and in some cases impossible for all practical purposes.

1. Symmetric-Key Encryption

With symmetric-key encryption, the encryption key can be calculated from the decryption key and vice versa. With most symmetric algorithms, the same key is used for both encryption and decryption, as shown in Figure 7-1.

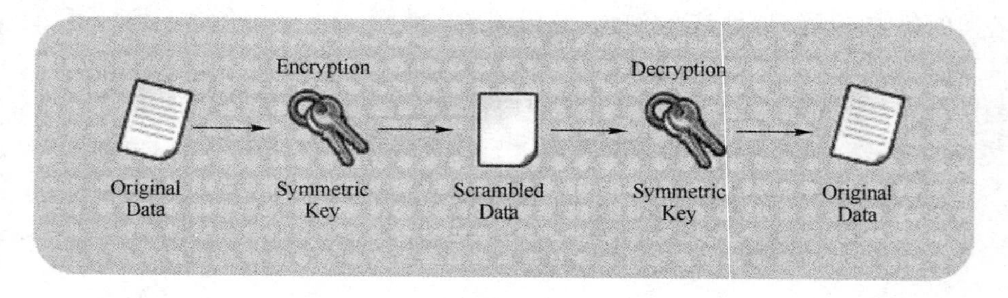

Figure 7-1 Diagram of Symmetric-Key Encryption

Implementations of symmetric-key encryption can be highly efficient, so that users do not experience any significant time delay as a result of the encryption and decryption. Symmetric-key encryption also provides a degree of authentication, since information encrypted with one symmetric key cannot be decrypted with any other symmetric key. Thus, as long as the symmetric key is kept secret by the two parties using it to encrypt communications, each party can be sure that it is communicating with the other as long as the decrypted messages continue to make sense.

Symmetric-key encryption is effective only if the symmetric key is kept secret by the two parties involved. If anyone else discovers the key, it affects both confidentiality and authentication. A person with an unauthorized symmetric key cannot only decrypt messages sent with that key, but can encrypt new messages and send them as if they came from one of the two parties who were originally using the key.

Symmetric-key encryption plays an important role in the SSL protocol, which is widely used for authentication, tamper detection, and encryption over TCP/IP networks. SSL also uses techniques of public-key encryption.

2. Public-Key Encryption

The most commonly used implementations of public-key encryption are based on algorithms patented by RSA Data Security. Therefore, this section describes the RSA[1] approach to public-key encryption.

Public-key encryption (also called asymmetric encryption) involves a pair of keys—a public key and a private key—associated with an entity that needs to authenticate its identity electronically or to sign or encrypt data. Each public key is published, and the corresponding private key is kept secret. Data encrypted with your public key can be decrypted only with your private key. Figure 7-2 shows a simplified view of the way public-key encryption works.

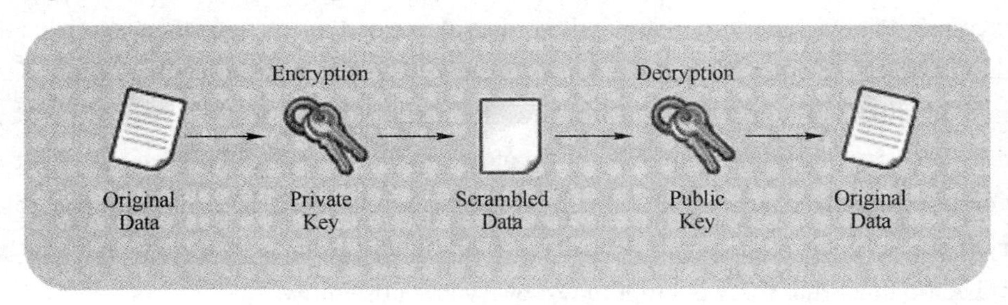

Figure 7-2 Diagram of Public-Key Encryption

The scheme shown in Figure 7-2 lets you freely distribute a public key, and only you will be able to read data encrypted using this key. In general, to send encrypted data to someone, you encrypt the data with that person's public key, and the person receiving the encrypted data decrypts it with the corresponding private key.

Compared with symmetric-key encryption, public-key encryption requires more computation and is therefore not always appropriate for large amounts of data. However, it's possible to use public-key encryption to send a symmetric key, which can then be used to encrypt additional data. This is the approach used by the SSL protocol.

As it happens, the reverse of the scheme shown in Figure 7-2 also works: data encrypted with your private key can be decrypted only with your public key. This would not be a desirable way to encrypt sensitive data, however, because it means that anyone with your public key, which is by definition published, could decrypt the data. Nevertheless, private-key encryption is useful, because it means you can use your private key to sign data with your digital signature—An important requirement for electronic commerce and other commercial applications of cryptography. Client software such as Firefox can then use your public key to confirm that the message was signed with your private key and that it hasn't been tampered with since being signed.

3. Key Length and Encryption Strength

Breaking an encryption algorithm is basically finding the key to the access the encrypted data in plain text. For symmetric algorithms, breaking the algorithm usually means trying to determine the key used to encrypt the text. For a public key algorithm, breaking the algorithm usually means acquiring the shared secret information between two recipients.

One method of breaking a symmetric algorithm is to simply try every key within the full algorithm until the right key is found. For public key algorithms, since half of the key pair is publicly known, the other half (private key) can be derived using published, though complex, mathematical calculations. Manually finding the key to break an algorithm is called a brute force attack.

Breaking an algorithm introduces the risk of intercepting, or even impersonating and fraudulently verifying private information.

The key strength of an algorithm is determined by finding the fastest method to break the algorithm and comparing it to a brute force attack.

For symmetric keys, encryption strength is often described in terms of the size or length of the keys used to perform the encryption: In general, longer keys provide stronger encryption. Key length is measured in bits. For example, 128-bit keys for use with the RC4[2] symmetric-key cipher supported by SSL provide significantly better cryptographic protection than 40-bit keys for use with the same cipher. Roughly speaking, 128-bit RC4 encryption is 3×1026 times stronger than 40-bit RC4 encryption. An encryption key is considered full strength if the best known attack to break the key is no faster than a brute force attempt to test every key possibility.

Different ciphers may require different key lengths to achieve the same level of encryption strength. The RSA cipher used for public-key encryption, for example, can use only a subset of all possible values for a key of a given length, due to the nature of the mathematical problem on which it is based. Other ciphers, such as those used for symmetric key encryption, can use all possible values for a key of a given length, rather than a subset of those values.

Because it is relatively trivial to break an RSA key, an RSA public-key encryption cipher must have a very long key, at least 1024 bits, to be considered cryptographically strong. On the other hand, symmetric-key ciphers can achieve approximately the same level of strength with an 80-bit key for most algorithms.

New Words

transform *v.* 转换，改变，改造，使……变形；转化，变换

unintelligible *adj.* 难解的，无法了解的，莫明其妙的

recipient　　*n.* 接受者，收件人
　　adj. 容易接受的，感受性强的
decryption　　*n.* 解密，译码
intelligible　　*adj.* 可理解的
cryptographic　　*adj.* 用密码写的，密码的
cipher　　*n.* 密码
　　v. 将……译为密码
decrypt　　*v.* 译码
symmetric　　*adj.* 对称的，均衡的
calculated　　*adj.* 计算出的
vice versa　　*adv.* 反之亦然
implementation　　*n.* 执行
experience　　*n.* 经验，体验，经历
significant　　*adj.* 有意义的，重大的，重要的
delay　　*v.* & *n.* 耽搁，延迟，迟滞
involved　　*adj.* 有关的
originally　　*adv.* 最初，原先
asymmetric　　*adj.* 不均匀的，不对称的
sign　　*v.* 在……上签名，在……上签字，签署
publish　　*v.* 公开，公布，发表
corresponding　　*adj.* 相应的
computation　　*n.* 计算，估计
desirable　　*adj.* 值得要的，合意的，令人想要的
sensitive　　*adj.* 敏感的，灵敏的
cryptography　　*n.* 密码系统，密码术
confirm　　*v.* 确定，批准，使有效
acquire　　*v.* 获得
impersonate　　*v.* 模仿，扮演，人格化，拟人
fraudulent　　*adj.* 欺诈的，欺骗性的
verify　　*v.* 检验，校验，查证，核实
measure　　*v.* 测量，度量，估量
　　n. 量度标准，测量，措施
bit　　*n.* 位，比特

Phrases

mathematical function　　数学函数
for all practical purposes　　实际上
symmetric-key encryption　　对称密钥加密

as long as	只要；在……的时候
communicate with	通话
play an important role in…	在……中起重要作用
public-key encryption	公钥加密
asymmetric encryption	非对称加密，不对称加密
a pair of	一双，一对
electronic commerce	电子商务
key length	钥匙长度
encryption strength	加密强度
encryption algorithm	加密算法
plain text	明文

Abbreviations

SSL (Security Socket Layer)	加密套接字层，安全套接字层

Notes

[1] RSA is algorithm used by modern computers to encrypt and decrypt messages. It is an asymmetric cryptographic algorithm. Asymmetric means that there are two different keys. This is also called public key cryptography, because one of them can be given to everyone. The other key must be kept private. It is based on the fact that finding the factors of an integer is hard (the factoring problem). RSA stands for Ron Rivest, Adi Shamir and Leonard Adleman, who first publicly described it in 1977. A user of RSA creates and then publishes the product of two large prime numbers, along with an auxiliary value, as their public key. The prime factors must be kept secret. Anyone can use the public key to encrypt a message, but with currently published methods, if the public key is large enough, only someone with knowledge of the prime factors can feasibly decode the message.

[2] RC4 (Rivest Cipher 4) is algorithm invented by Ronald Rivest in 1987. In cryptography, RC4 is a stream cipher. While remarkable for its simplicity and speed in software, multiple vulnerabilities have been discovered in RC4, rendering it insecure. It is especially vulnerable when the beginning of the output keystream is not discarded, or when nonrandom or related keys are used. Particularly problematic uses of RC4 have led to very insecure protocols such as WEP.

Exercises

[Ex. 1] Answer the following questions according to the text.

1. What is encryption?
2. What is decryption?
3. What is a cryptographic algorithm?
4. What is the ability to keep encrypted information secret with most modern cryptography?

5. When is symmetric-key encryption effective? What happens if anyone else discovers the key?

6. What are the most commonly used implementations of public-key encryption based on?

7. Why is private-key encryption useful?

8. What does breaking the algorithm usually mean for symmetric algorithms and a public key algorithm?

9. For symmetric keys, how is encryption strength often described? How is key length measured?

10. Why must an RSA public-key encryption cipher have a very long key, at least 1024 bits, to be considered cryptographically strong?

[Ex. 2] Translate the following terms or phrases from English into Chinese and vice versa.

1. asymmetric encryption	1.
2. key length	2.
3. plain text	3.
4. public-key encryption	4.
5. symmetric-key encryption	5.
6. 确定，批准，使有效	6.
7. 不均匀的，不对称的	7.
8. 用密码写的，密码的	8.
9. 解密，译码	9.
10.检验，校验，查证，核实	10.

[Ex. 3] Translate the following passage into Chinese.

What Is an RSA Algorithm?

RSA Algorithm is named after its joint inventors, Ron Rivset, Adi Shamir and Leonard Adleman who invented it in 1977. This algorithm is the first of its kind that can be used for public-key encryption as well as digital signatures. The RSA Algorithm is extremely secure due to long keys and up-to-date implementations. It is widely used in e-commerce protocols. An algorithm is a process that produces a particular type of encryption in cryptography known as "encryption algorithm". Encryption allows safe message transmission over the Internet.

1. What are the different steps for RSA algorithm?

The RSA algorithm involves three main steps: key generation, encryption and decryption.

2. What is key generation?

It involves the generation of a public key and a private key. The public key is known to everyone. It is specifically used to decipher messages. Private key is used to decrypt the messages that are encrypted using the public key. The message or the data is encrypted using the public key. Encryption involves converting the message into a coded format. The public key is accessible to everyone and anyone can encrypt the message.

3. What is decryption?

The intended recipient of the message has a private key. The user uses this private key to decrypt the encrypted message and reads the message. The private key has to be kept confidential.

4. What is a digital signature?

Let us consider an example: A sends an encrypted message using B's public key. In the message, A claims to be the sender but B has no means of verifying that the message was actually from A as anyone has access to B's public key. In order to verify, RSA Algorithm is used to digitally sign the message. Now, A sends a signed message to B by using her own private key. A can produce a hash value while decrypting the message and attach it as a signature with the message. When B receives the signed message, he uses the same hash algorithm with A's public key and compares the resulting hash value with the actual hash value of the message. If the two values match, B is assured that A is the actual sender of the message.

[Ex. 4] Fill in the blanks with the words given below.

companies	advanced	provide	break	meet
algorithm	encryption	standard	method	special

What Is DES Cryptography?

Cryptography is the process of creating secret messages in an effort to hide sensitive data. In computer science, there are many methods of encrypting data. The Data Encryption Standard (DES) was the first defined standard for computer ___1___. DES cryptography was created in 1976 by a group from IBM. At that time it was considered the standard ___2___ for creating encrypted data for the United States' government.

DES cryptography is based on a special 56-bit encryption key ___3___. The encryption key is the primary method for encrypting and decrypting messages. This encryption process is typically referred to as ciphering and deciphering secret messages. Encryption is a process of converting simple strings of text data into a scrambled version of characters. This cryptographic process is completed by using ___4___ hashing algorithms with the unique 56-bit encryption key.

The National Institute of Standards and Technology (NIST) is the governing body that manages encryption standards within the United States. This group accepted DES cryptography as the defined ___5___ for data encryption between the years of 1974 to 2001 for all government agencies. In 2001 DES cryptography was superseded by the Advanced Encryption Standard (AES). The new standard supports a more ___6___encryption key of 256-bit.

There were many permutations of the DES cryptography during its reign as the standard for data encryption. In early 1986 it was used in video ciphering. This encryption process was the defined method used by cable ___7___ to scramble cable video broadcasts. This forced customers to purchase special video cable boxes that included the DES encryption algorithm. This algorithm was required to unscramble the video broadcast.

The primary issue with DES cryptography is the size of the encryption key. The 56-bit key did not ___8___ enough of a deterrent for computer hackers. The DES standard was quickly deciphered and many black-market encryption algorithms became readily available.

In 1998 an encryption program was created to prove the weakness of the DES cryptography. This was created by the Electronic Frontier Foundation (EFF) and known as DES cracker called "Deep Crack". The program was able to ___9___ the code for DES in 56 hours. This was the final blow to the DES standard, which forced the creation of the new AES standard.

AES cryptography was declared the standard encryption by NIST in 2001. Today there are many encryption algorithms available that ___10___ the AES standard. Most of these algorithms offer an extremely high level of security that cannot be cracked. DES cryptography is typically only supported in legacy systems that cannot support a large encryption keys.

Text B

How Does Encryption Work, and Is It Really Safe?

For many, the word "encryption" probably stirs up James Bond-esque images of a villain with a briefcase handcuffed to his wrist with nuclear launch codes or some other action movie staple. In reality, we all use encryption technology on a daily basis, and while most of us probably don't understand the "how" or the "why", we are sure that data security is important, and if encryption helps us to accomplish that, then we're definitely on board.

Nearly every computing device we interact with on a daily basis utilizes some form of encryption technology. From smartphones (which can often have their data encrypted), to tablets, desktop, laptops or even your trusty Kindle, encryption is everywhere.

1. What is Encryption?

Encryption is a modern form of cryptography that allows a user to hide information from others. Encryption uses a complex algorithm called a cipher in order to turn normalized data (plaintext) into a series of seemingly random characters (ciphertext) that is unreadable by those without a special key in which to decrypt it. Those that possess the key can decrypt the data in order to view the plaintext again rather than the random character string of ciphertext.

Two of the most widely used encryption methods are public key (asymmetric) encryption and private key (symmetric) encryption. The two are similar in the sense that they both allow a user to encrypt data to hide it from others, and then decrypt it in order to access the original plaintext. They differ, however, in how they handle the steps between encryption and decryption.

2. Public-Key Encryption

Public-key or asymmetric encryption uses the recipient's public key as well as a (mathematically) matching private key, see Figure 7-3.

Figure 7-3 Diagram of Public Key Encryption Process

For example, if Joe and Karen both had keys to a box, with Joe having the public key and Karen having a matching private key, Joe could use his key to unlock the box and put things into it, but he wouldn't be able to view items already in there, nor would he be able to retrieve anything. Karen, on the other hand, could open the box and view all items inside as well as removing them as she saw fit by using her matching private key. She could not, however, add things to the box without having an additional public key.

In a digital sense, Joe can encrypt plaintext (with his public key), and send it to Karen, but only Karen (and her matching private key) could decrypt the ciphertext back into plaintext. The public key (in this scenario) is used for encrypting ciphertext, while the private key is used to decrypt it back into plaintext. Karen would only need the private key to decrypt Joe's message, but she'd need access to an additional public key in order to encrypt a message and send it back to Joe. Joe on the other hand couldn't decrypt the data with his public key, but he could use it to send Karen an encrypted message.

3. Private-Key Encryption

Where private-key or symmetric encryption differs from public-key encryption is in the purpose of the keys themselves. There are still two keys needed to communicate, but each of these keys is now essentially the same, see Figure 7-4.

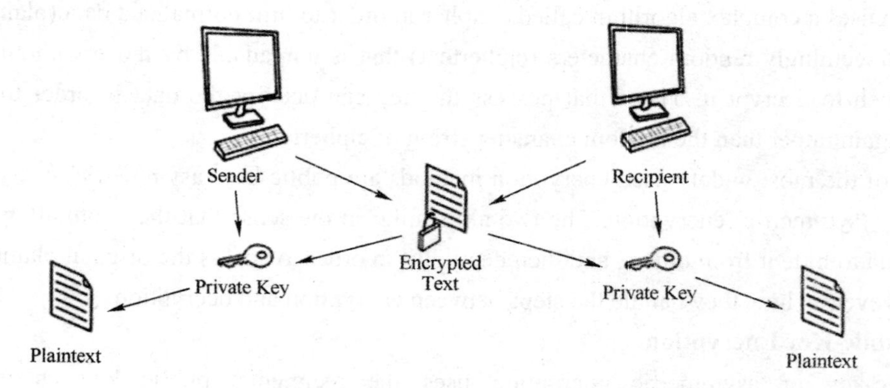

Figure 7-4 Diagram of Private Key Encryption Process

For example, Joe and Karen both possess keys to the aforementioned box, but in this scenario the keys do the same thing. Both of them are now able to add or remove things from the box.

Speaking digitally, Joe can now encrypt a message as well as decrypt it with his key. Karen can do the same with hers.

4. A Brief History of Encryption

When talking about encryption, it's important to make the distinction that all modern encryption technology is derived from cryptography. Cryptography is—at its core—the act of creating and (attempting to) decipher a code. While electronic encryption is relatively new in the grander scheme of things, cryptography is a science that dates back to ancient Greece.

The Greeks weren't alone in developing primitive cryptography methods. The Romans followed suit by introducing what came to be known as "Caesar's cipher", a substitution cipher that involved substituting a letter for another letter shifted further down the alphabet. For example, if the key involved a right shift of three, the letter A would become D, the letter B would be E, and so on.

Enigma machine: The Enigma machine is a WWII technology known as an electromechanical rotor cipher machine. This device looked like an oversized typewriter and allowed operators to type in plaintext, while the machine encrypted the message and sent it to another unit. The receiver wrote down the random string of encrypted letters after they lit up on the receiving machine and broke the code after setting up the original pattern from the sender on his machine.

Data Encryption Standard: The Data Encryption Standard (DES) was the first modern symmetric key algorithm used for encryption of digital data. Developed in the 1970s at IBM, DES became the Federal Information Processing Standard for the United States in 1977 and became the foundation for which modern encryption technologies were built.

5. Modern Encryption Technology

Modern encryption technology uses more sophisticated algorithms as well as larger key sizes in order to better conceal encrypted data. As key size continues to improve, the length of time it takes to crack an encryption using a brute force attack skyrockets. For example, while a 56-bit key and a 64-bit key look to be relatively close in value, the 64-bit key is actually 256 times harder to crack than the 56-bit key. Most modern encryptions use a minimum of a 128-bit key, with some using 256-bit keys or greater. To put that into perspective, cracking a 128-bit key would require a brute force attack to test over 339,000,000,000,000,000,000,000,000,000,000,000 possible key combinations. In case you are curious, it would actually take over a million years to guess the correct key using brute force attacks, and that's using the most powerful supercomputers in existence. In short, it's theoretically implausible that anyone would even try to break your encryption using 128-bit or higher technology.

(1) 3 DES

Encryption standards have come a long way since DES was first adopted in 1977. In fact, a new DES technology, known as Triple DES (3DES) is quite popular, and it's based on a modernized version of the original DES algorithm. While the original DES technology was rather limited with a key size of just 56 bits, the current 3DES key size of 168 bits make it significantly more difficult and time consuming to crack.

(2) AES

The Advanced Encryption Standard (AES) is a symmetric cipher based on the Rijandael block cipher that is currently the United States federal government standard. AES was adopted worldwide as the heir apparent to the now deprecated DES standard of 1977 and although there are published examples of attacks that are faster than brute force, the powerful AES technology is still thought to be computationally infeasible in terms of cracking. In addition, AES offers solid performance on a wide variety of hardware and offers both high speed and low RAM requirements making it a top-notch choice for most applications. If you're using a Mac, the popular encryption tool FileVault is one of many applications that uses AES.

(3) RSA

RSA is one of the first widely used asymmetric cryptosystems for data transmission. The algorithm was first described in 1977, and relies on a public key based on two large prime numbers and an auxiliary value in order to encrypt a message. Anyone can use the public key in order to encrypt a message but only someone with knowledge of the prime numbers can feasibly attempt to decode the message. RSA opened the doors to several cryptographic protocols such as digital signatures and cryptographic voting methods. It's also the algorithm behind several open source technologies, such as PGP[1], which allows you to encrypt digital correspondence.

(4) ECC[2]

Elliptic curve cryptography is among the most powerful and least understood forms of encryption used today. Proponents of the ECC approach cite the same level of security with faster operational time largely due to the same levels of security while utilizing smaller key sizes. The high performance standards are due to the overall efficiency of the elliptic curve, which makes them ideal for small embedded systems such as smart cards. The NSA is the biggest supporter of the technology, and it's already being billed as the successor to the aforementioned RSA approach.

6. So, is encryption safe?

Unequivocally, the answer is yes. The amount of time, energy usage and computational cost to crack most modern cryptographic technologies makes the act of attempting to break an encryption (without the key) an expensive exercise that is, relatively speaking, futile. That said, encryption does have vulnerabilities that rest largely outside of the power of the technology.

For example:

(1) Backdoors

No matter how secure the encryption, a backdoor could potentially provide access to the private key. This access provides the means necessary to decrypt the message without ever breaking the encryption.

(2) Private Key Handling

While modern encryption technology is extremely secure, humans aren't as easy to count on. An error in handling the key such as exposure to outside parties due to a lost or stolen key, or human error in storing the key in insecure locations could give others access to encrypted data.

(3) Increased Computational Power

Using current estimates, modern encryption keys are computationally infeasible to crack. That said, as processing power increases, encryption technology needs to keep pace in order to stay ahead of the curve.

(4) Government Pressure

The legality of this is dependent on your home country, but typically speaking, mandatory decryption laws force the owner of encrypted data to surrender the key to law enforcement personnel (with a warrant / court order) to avoid further prosecution. In some countries, such as Belgium, owners of encrypted data that are concerned with self-incrimination aren't required to comply, and police are only allowed to request compliance rather than demand it. Let's not forget, there is also precedent of website owners willfully handing over encryption keys that stored customer data or communications to law enforcement officials in an attempt to remain cooperative.

Encryption isn't bulletproof, but it protects each and every one of us in just about every aspect of our digital lives. While there is still a (relatively) small demographic that doesn't trust online banking or making purchases at Amazon or other online retailers, the rest of us are quite a bit safer shopping online (from trusted sources) than we would be from taking that same shopping trip at our local mall.

While your average person is going to remain blissfully unaware of the technologies protecting them while purchasing coffee at Starbucks with their credit card, or logging on to Facebook, that just speaks to the power of the technology. You see, while the tech that we get excited about is decidedly more sexy, it's those that remain relatively unseen that are doing the greatest good. Encryption falls firmly into this camp.

New Words

image	*n.*	形象，概念，声誉
villain	*n.*	坏人
handcuff	*v.*	给……戴上手铐；限制
	n.	手铐
staple	*n.*	主要部分，重要内容
accomplish	*v.*	完成，达到，实现
utilize	*v.*	利用
desktop	*n.*	桌上型计算机
laptop	*n.*	膝上型计算机，笔记本式计算机
normalize	*v.*	使正常化，使标准化，规格化
random	*n.*	随机，任意
	adj.	随机的，任意的
character	*n.*	字符
ciphertext	*n.*	密文，暗文
aforementioned	*adj.*	上述的，前述的

distinction	*n.*	区别，差别
decipher	*v.*	破译密码，解读
	n.	密电译文
substitution	*n.*	代替，代入法，置换
electromechanical	*adj.*	机电的，电机的
rotor	*n.*	转子，回转轴，转动体
sophisticate	*v.*	使复杂化，使……精巧
	n.	久经世故的人，精于……之道的人
skyrocket	*v.*	突升，猛涨
supercomputer	*n.*	超级计算机
implausible	*adj.*	难以置信的
modernize	*v.*	使现代化
adopt	*v.*	采用
deprecate	*v.*	不赞成，抨击，反对；藐视，轻视
computational	*adj.*	计算的
infeasible	*adj.*	不可实行的
top-notch	*adj.*	拔尖的
cryptosystem	*n.*	密码系统
auxiliary	*adj.*	辅助的
decode	*v.*	解码，译码
correspondence	*n.*	通信
proponent	*n.*	建议者，支持者
successor	*n.*	继承者，接任者，后续的事物
unequivocal	*adj.*	不含糊的
expensive	*adj.*	花费的，昂贵的
futile	*adj.*	无用的，无效果的
surrender	*v.*	交出
warrant	*n.*	授权，正当理由，批准
	v.	保证，批准，使有正当理由
prosecution	*n.*	检举，起诉
self-incrimination	*n.*	自证其罪
bulletproof	*adj.*	防弹的
demographic	*adj.*	人口统计学的
sexy	*adj.*	时髦的，性感的，诱人的，迷人的

Phrases

stir up	激起
nuclear launch code	核弹发射代码
be on board	在船（火车，飞机，汽车）上

interact with…	与……相合
hide from	隐瞒
on the other hand	另一方面
follow suit	跟着做
Enigma machine	恩尼格玛机
Triple DES (3DES)	三重数据加密算法
time consuming	耗费时间的，旷日持久的
data transmission	数据传输
embedded system	嵌入系统
count on	依靠，指望
be dependent on	依靠，依赖
court order	法院指令，法庭庭谕
be concerned with	参与，干预
online retailer	在线零售商

Abbreviations

WWII (World War II)	第二次世界大战
DES (Data Encryption Standard)	数据加密标准
AES (Advanced Encryption Standard)	高级加密标准
PGP (Pretty Good Privacy)	完美隐私，良好隐私密码法
ECC (Elliptic Curve Cryptography)	椭圆曲线密码
NSA (National Security Agency)	（美国）国家安全局

Notes

[1] PGP encryption program provides cryptographic privacy and authentication for data communication. PGP is used for signing, encrypting, and decrypting texts, e-mails, files, directories, and whole disk partitions and to increase the security of e-mail communications.

[2] Elliptic Curve Cryptography (ECC) is an approach to public-key cryptography based on the algebraic structure of elliptic curves over finite fields. ECC requires smaller keys compared to non-ECC cryptography (based on plain Galois fields) to provide equivalent security.

Elliptic curves are applicable for key agreement, digital signatures, pseudo-random generators and other tasks. Indirectly, they can be used for encryption by combining the key agreement with a symmetric encryption scheme. They are also used in several integer factorization algorithms based on elliptic curves that have applications in cryptography, such as Lenstra elliptic curve factorization.

Exercises

[Ex. 5] Answer the following questions according to the text.

1. What is encrytion?
2. What are two of the most widely used encryption methods mentioned in the text?

3. What is cryptography?

4. What is the Enigma machine?

5. What does DES stand for? What was it?

6. What does modern encryption technology use in order to better conceal encrypted data?

7. When was DES first adopted? What is a new DES technology known as?

8. What is the Advanced Encryption Standard?

9. What is RSA?

10. What is among the most powerful and least understood forms of encryption used today?

Reading Material

Understanding Encryption and Password Protection

Encryption is critically important to any sort of security because a file that only has a password[1] on it and is not encrypted can be opened and read by anyone with the right skills. That is worse than no security at all, because it gives a false impression[2] of security. When you apply a password to a file in Attach Plus, the contents of the file are actually encrypted. Without the password, the decryption keys are not known, so the content cannot be recovered by an attacker.

There are several key concepts to understand regarding encryption:

1) The two classes of encryption algorithms;

2) Strategies for key exchange;

3) How encryption and decryption keys are generated in symmetric algorithms;

4) The importance of using good password.

1. Two Classes of Encryption Algorithms

Encryption algorithms take a plain text stream of data and an encryption key and generate a cipher text stream of data. There are two broad classifications of encryption algorithms, split by whether they use the same key for encryption as for decryption.

1) Symmetric—this means that the same encryption key is used for decryption.

2) Asymmetric—this means that there are "different" keys for encryption and decryption.

In general, symmetric encryption algorithms are fast. Asymmetric algorithms are very slow. This has no bearing on[3] the relative "security" of either class of algorithms—just the speed with which the algorithm can be executed.

The most widely used symmetric encryption algorithms are 3DES and AES. These use shared[4] keys, and are actually the algorithms responsible for[5] the vast bulk of[6] data transferred securely

---

1 password *n.* 密码，口令

2 impression *n.* 印象，感想

3 have no bearing on 与……无关，对……没有影响

4 shared *adj.* 共享的

5 be responsible for 为……负责，是造成……的原因

6 the bulk of 大半，大部，大多数

over the Internet.

The most widely used asymmetric encryption algorithm is referred to as "Public Key". The idea here is pretty interesting—your recipient can actually publish their encryption key, and anyone can use that key to encrypt data. But only the recipient knows the private decryption key. This sounds ideal, except for two glaring[1] limitations:

1) Asymmetric encryption is slow. This issue, as it turns out, is easy to fix. You use the recipient's public key and asymmetric encryption to encrypt a randomly generated "symmetric" key. You send the encrypted package to the recipient. The recipient is able to decrypt the package (because they have the private key), and obtain the symmetric key. You then do all of your exchange using symmetric algorithms, which are much faster. This strategy is used quite frequently. In fact, most secure exchanges of data over the Internet begin with a single asymmetric exchange to transfer temporary[2] keys, followed by the symmetric exchange of the actual data.

But there's a big remaining problem:

2) The recipient has to have a public key. In a situation where two large businesses are exchanging data (think of an ATM communicating with the bank's servers), the cost of setting up what is referred to as "Public Key Infrastructure" (or PKI) is not a deterrent.

They pay money to have certificates generated and authenticated. They set up servers to publish these keys, etc. With smaller businesses, setting up public key certificates is still within reach (most firms have some sort of certificate for use with their website), but it does add a lot of complexity. The security of the private key must also be managed. Compromising a private key is horrendous[3] from a security perspective. Making sure that you don't accidentally write a private key to a disk somewhere is incredibly important. In fact, most schemes involve encrypting the private key with—that's right—a password using symmetric encryption.

When you are dealing with end user recipients, this problem becomes untenable[4]. It would be convenient[5] (though a bit intrusive) if every American was issued a public/private key at birth, and it was maintained in a central repository[6]. But that just isn't the case.

There are services that will issue public/private keys, but this is a huge technological barrier[7] to entry for the vast majority of people you might want to communicate with.

2. Key Exchange

Exchange of encryption and/or decryption keys is probably one of the most difficult aspects of cryptography. At a purely technical level, symmetric/asymmetric hybrid[8] encryption schemes

---

1 glaring *adj.* 耀眼的，显眼的

2 temporary *adj.* 暂时的，临时的

3 horrendous *adj.* 可怕的

4 untenable *adj.* 防守不住的，不能维持的，支持不住的

5 convenient *adj.* 便利的，方便的

6 repository *n.* 贮藏室，仓库

7 barrier *n.* 障碍物，栅栏，屏障

8 hybrid *n.* 混合物

are appealing[1] from a key exchange perspective, but they have real world limitations that make them unrealistic[2] for use in business to customer communication.

There are also distinct advantages to pure symmetric encryption. Specifically, the question of key exchange is completely deferred. It takes place with a phone call, or face-to-face meeting. Central repositories of keys don't need to be created, managed, paid for, etc. The end recipient is probably going to have to use a password to protect their private keys regardless, so from their perspective, they still have to type the password in.

The question of key exchange is actually where the Attach Plus design began. We wanted to provide a solution that was incredibly simple for the sender and the recipient. We needed this to be part of the process of actually attaching files to[3] e-mails (trying to get people to context switch to handle encryption was laughable). In addition, we needed the application to "just work", regardless of whom your recipient is, and preferably without the recipient having to install anything on their computers, or sign up for any special accounts, etc.

3. Encryption Strength and How Keys Are Generated in Symmetric Algorithms

No discussion of encryption would be complete without talking about encryption strength. Interestingly, a well designed encryption algorithm is, itself, un-crackable without the decryption key. But we can guess the decryption key and just try them all, right? So the best encryption algorithm on the planet is useless if it is protected by a 3 digit number—a computer can try 999 different combinations very quickly. So once we've cleared the hurdle[4] that a particular algorithm is well designed, we have to ensure that the size of the encryption key is big enough that it would take an absurdly long time for an attacker to guess every single combination.

The size of the encryption key is usually measured in "bits". Most encrypted traffic that goes over the Internet uses 128-bit encryption. Generally speaking, the larger the key, the stronger the encryption—but you have to be a bit careful with this. A modern computer would require many, many thousands of years to try every combination of a 128-bit encryption key (this is called a "brute force" attack). By going to 256-bit encryption, you can change that to millions and millions of years. But the "effective" security is the same. 128-bit key lengths have been found by the security community to provide a nice balance[5] of encryption that would take a really, really long time to crack with brute force techniques, without the key becoming ridiculously long.

When you supply a password to Attach Plus, the software takes the password and turns it into a 128 (or 256) bit key using an algorithm called a one-way hashing algorithm. If you input the same password, you get the same 128 or 256-bits of effectively random data. But it is not possible to start with the 128-bits of data and obtain the password.

The files generated by Attach Plus are actually not password protected. In fact, the password

---

1 appealing *adj.* 吸引人的

2 unrealistic *adj.* 不切实际的，不实在的

3 attach to 把……放在，把……附加到

4 hurdle *n.* 篱笆，栏，障碍

5 balance *n.* 平衡

isn't part of the file at all. Instead, the contents are encrypted using a symmetric algorithm. The key for the symmetric algorithm is obtained by taking the password and processing it with a oneway hashing algorithm.

When a user opens the attachment, they are prompted for a password. The password is not compared against any value in the attachment. The password is used to mathematically obtain the decryption key.

4. The Importance of Using Good Passwords

Let's start with examples of bad passwords:

1) Regular English words (i.e. not including numbers, symbols[1], combination of upper and lower case[2]).

2) A client's social security number, or any other piece of information that might be guessable[3].

3) Any password that you e-mail to the recipient (if an attacker has access to one e-mail, they probably have access to all of them!).

4) Anything that is short (shorter than 8 characters is generally considered to be bad). My recommendation[4] is to let the recipient decide the password. That puts the onus [5]on them to choose something that is secure.

By analogy: The strongest safe in the world does no good if the combination is written on the door.

参 考 译 文

加密和解密

加密是转换信息的过程，使得除了预期的收件人之外任何人都不能理解该信息。解密是将已经加密的信息转换为可以再次理解的信息的过程。密码算法也称译码，是用于加密或解密的数学函数。在大多数情况下，采用两个相关函数，一个用于加密，另一个用于解密。

对于大多数现代密码学，保持加密信息安全的能力不是基于广为人知的加密算法，而是基于被称为密钥的数字，该密钥必须与算法一起使用以产生加密结果或解密先前加密的信息。使用正确的密钥进行解密很简单。没有正确密钥的解密是非常困难的，在某些情况下，实际上是不可能的。

1. 对称密钥加密

使用对称密钥加密，可以从解密密钥计算出加密密钥，反之亦然。对于大多数对称算法，加密和解密都使用相同的密钥，如图 7-1 所示。

使用对称密钥加密效率很高，这样用户不会因为加密和解密而经历任何显著的时间延

1 symbol *n.* 符号

2 upper and lower case 大写和小写

3 guessable *adj.* 可猜测的，可推测的

4 recommendation *n.* 推荐，劝告，建议

5 onus *n.* 责任，负担

迟。对称密钥加密还提供一定程度的认证，因为用一个对称密钥加密的信息不能用任何其他对称密钥解密。因此，只要双方使用对称密钥来加密通信时不泄露对称密钥，在解密的消息继续有意义时，每一方都可确认正在与对方通信。

对称密钥加密只有当事双方在对称密钥保密的情况下才有效。如果别人发现了密钥，那么就会影响机密性和真实性。未经授权却具有对称密钥的人不仅可以解密用该密钥发送的消息，而且还可以对新消息进行加密，并将它们发送出去，就像他们来自最初使用密钥的双方之一那样。

对称密钥加密在 SSL 协议中起着重要的作用，SSL 协议广泛用于通过 TCP / IP 网络的认证、篡改检测和加密。SSL 还使用公钥加密技术。

2. 公钥加密

公钥加密最常用的实现方法是基于有专利权的 RSA Data Security 算法。因此，本节介绍公钥加密的 RSA 方法。

公钥加密（也称非对称加密）涉及一对密钥——一个公钥和一个私钥，与一个需要以电子方式对其进行身份验证或签名或加密数据的实体相关。每个公钥公开而对应的私钥保密。使用公钥加密的数据只能使用你的私钥进行解密。图 7-2 所示为公开密钥加密方式的简化视图。

图 7-2 所示的方案允许你自由分发公钥，只有你可以读取使用此密钥加密的数据。一般来说，要将加密数据发送给某人，请使用该人的公钥加密数据，接收加密数据的人员将使用相应的私钥对其进行解密。

与对称密钥加密相比，公钥加密需要更多的计算，因此并不总是适合大量的数据。然而，可以使用公钥加密来发送一个对称密钥，然后可以用它来加密附加数据。这是 SSL 协议使用的方法。

当然，图 7-2 所示的方案也可以反向使用：使用你的私钥加密的数据只能使用你的公钥解密。然而，这不是加密敏感数据的理想方式，因为这意味着任何拥有你的公钥（按照定义发布）的人都可以对数据进行解密。然而，私钥加密是有用的，因为这意味着你可以使用你的私钥以数字签名的形式来签署数据，这是电子商务和其他商业密码学应用的重要要求。客户端软件（如 Firefox）可以使用你的公钥来确认该邮件已使用你的私钥签名，并且自从签名以来尚未被篡改。

3. 密钥长度和加密强度

破解加密算法基本上是找到一个密钥，以便以明文形式访问加密数据。对于对称算法，破解算法通常意味着尝试确定用于加密文本的密钥。对于公钥算法，破解算法通常意味着获取两个收件人之间的共享秘密信息。

破解对称算法的一种方法是简单地尝试完整算法中的每个密钥，直到找到正确的密钥。对于公钥算法，由于密钥对的一半是公知的，所以另一半（私钥）可以使用已发布的复杂的数学计算得到。手动找到破解算法的密钥称为暴力攻击。

破解了一个算法，就可以带来一下风险：窃取私有信息，或者甚至冒充他人，以及在验证私人信息时实施欺骗。

算法的密钥强度取决于找到最快的方法来破解算法的难易程度，并将其与暴力攻击进行比较。

对于对称密钥，通常根据用于执行加密的密钥的大小或长度来描述加密强度：较长的密钥一般提供更强的加密。钥匙长度以“位”为单位。例如，与 SSL 支持的 RC4 对称密钥密码一起使用的 128 位密钥提供比使用相同密码的 40 位密钥更好的加密保护。大致来说，128 位 RC4 加密比 40 位 RC4 加密强 3×1026 倍。如果攻破密钥的最佳已知攻击速度还不如强力测试每个钥匙的速度快，则加密密钥被认为是全强度的。

不同的密码可能需要不同的密钥长度来实现相同级别的加密强度。例如，用于公共密钥加密的 RSA 密码可以仅使用给定长度的密钥的所有可能值的子集，这是由于其所基于的数学问题的性质。其他密码（如用于对称密钥加密的密码）可以使用给定长度的密钥的所有可能值，而不是这些值的子集。

因为破解 RSA 密钥相当简单，RSA 公钥加密密码必须具有非常长的密钥，至少 1024 位才被认为是够加密强度的。另一方面，大多数算法使用 80 位密钥的对称密钥密码就可以实现大致相同的强度级别。

Unit 8

Text A

Public Key Infrastructure

A Public Key Infrastructure (PKI) is a set of roles, policies, and procedures needed to create, manage, distribute, use, store, and revoke digital certificates and manage public-key encryption. The purpose of a PKI is to facilitate the secure electronic transfer of information for a range of network activities such as ecommerce, Internet banking and confidential e-mail. It is required for activities where simple passwords are an inadequate authentication method and more rigorous proof is required to confirm the identity of the parties involved in the communication and to validate the information being transferred.

In cryptography, a PKI is an arrangement that binds public keys with respective identities of entities (like persons and organizations). The binding is established through a process of registration and issuance of certificates at and by a Certificate Authority (CA). Depending on the assurance level of the binding, this may be carried out by an automated process or under human supervision (see Figure 8-1) .

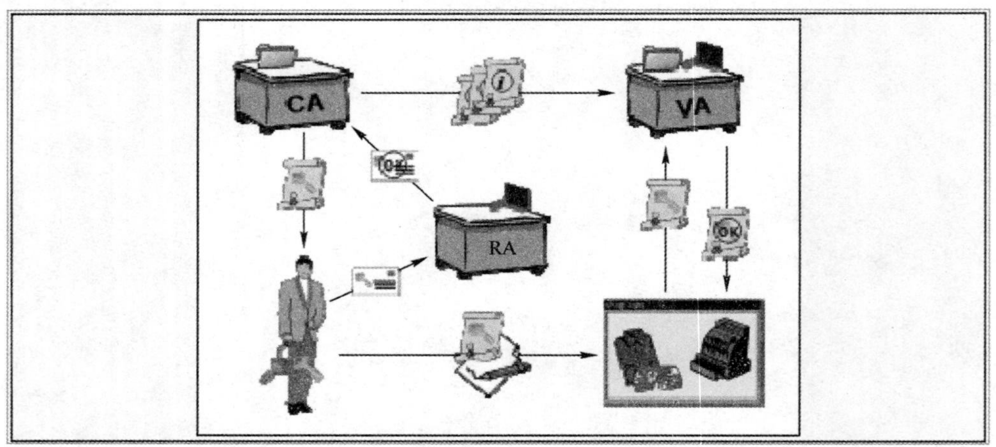

Figure 8-1 Diagram of a Public Key Infrastructure

The PKI role that assures valid and correct registration is called a Registration Authority (RA).

An RA is responsible for accepting requests for digital certificates and authenticating the entity making the request. In a Microsoft PKI, a registration authority is usually called a subordinate CA.

An entity must be uniquely identifiable within each CA domain on the basis of information about that entity. A third-party Validation Authority (VA)[1] can provide this entity information on behalf of the CA.

1. Design

Public key cryptography is a cryptographic technique that enables entities to securely communicate on an insecure public network, and reliably verify the identity of an entity via digital signatures.

A PKI is a system for the creation, storage, and distribution of digital certificates which are used to verify that a particular public key belongs to a certain entity. The PKI creates digital certificates which map public keys to entities, securely stores these certificates in a central repository and revokes them if needed.

A PKI consists of:

- A CA that stores, issues and signs the digital certificates.
- A registration authority which verifies the identity of entities requesting their digital certificates to be stored at the CA.
- A central directory—i.e., a secure location in which to store and index keys.
- A certificate management system managing things like the access to stored certificates or the delivery of the certificates to be issued.
- A certificate policy.

2. Methods of Certification

Broadly speaking, there have traditionally been three approaches to getting this trust: Certificate Authorities (CA), Web of Trust (WoT)[2], and Simple Public Key Infrastructure (SPKI).

(1) Certificate Authorities

The primary role of the CA is to digitally sign and publish the public key bound to a given user. This is done using the CA's own private key, so that trust in the user key relies on one's trust in the validity of the CA's key. When the CA is a third party separate from the user and the system, then it is called the Registration Authority (RA), which may or may not be separate from the CA. The key-to-user binding is established, depending on the level of assurance the binding has, by software or under human supervision.

The term Trusted Third Party (TTP)[3] may also be used for CA. Moreover, PKI is itself often used as a synonym for a CA implementation.

1) Issuer market share. In this model of trust relationships, a CA is a trusted third party—trusted both by the subject (owner) of the certificate and by the party relying upon the certificate.

According to NetCraft, although the global ecosystem is competitive, it is dominated by a handful of major CAs. Three certificate authorities (Symantec, Comodo, GoDaddy) account for three-quarters of all issued certificates on public-facing web servers.

2) Temporary certificates and single sign-on. This approach involves a server that acts as an

offline certificate authority within a single sign-on system. A single sign-on server will issue digital certificates into the client system, but never stores them. Users can execute programs, etc. with the temporary certificate. It is common to find this solution with X.509[4]-based certificates.

(2) Web of Trust

An alternative approach to the problem of public authentication of public key information is the web-of-trust scheme, which uses self-signed certificates and third party attestations of those certificates. The singular term "web of trust" does not imply the existence of a single web of trust, or common point of trust, but rather one of any number of potentially disjoint "webs of trust". Examples of implementations of this approach are PGP (Pretty Good Privacy) and GnuPG (an implementation of OpenPGP, the standardized specification of PGP). Because PGP and implementations allow the use of e-mail digital signatures for self-publication of public key information, it is relatively easy to implement one's own web of trust.

One of the benefits of the web of trust, such as in PGP, is that it can interoperate with a PKI CA fully trusted by all parties in a domain (such as an internal CA in a company) that is willing to guarantee certificates, as a trusted introducer. If the "web of trust" is completely trusted then, because of the nature of a web of trust, trusting one certificate is granting trust to all the certificates in that web.

The web of trust concept was first put forth by PGP creator Phil Zimmermann[5] in 1992 in the manual for PGP version 2.0. As time goes on, you will accumulate keys from other people that you may want to designate as trusted introducers. Everyone else will each choose their own trusted introducers. And everyone will gradually accumulate and distribute with their key a collection of certifying signatures from other people, with the expectation that anyone receiving it will trust at least one or two of the signatures. This will cause the emergence of a decentralized fault-tolerant web of confidence for all public keys.

(3) Simple Public Key Infrastructure

Another alternative, which does not deal with public authentication of public key information, is the Simple Public Key Infrastructure (SPKI) that grew out of three independent efforts to overcome the complexities of X.509 and PGP's web of trust. SPKI does not associate users with persons, since the key is what is trusted, rather than the person. SPKI does not use any notion of trust, as the verifier is also the issuer. This is called an "authorization loop" in SPKI terminology, where authorization is integral to its design.

(4) Blockchain-Based PKI

An emerging approach for PKI is to use the blockchain[6] technology commonly associated with modern cryptocurrency. Since blockchain technology aims to provide a distributed and unalterable ledger of information, it has qualities considered highly suitable for the storage and management of public keys. Emercoin is an example of a blockchain-based cryptocurrency that supports the storage of different public key types (SSH, GPG, RFC 2230, etc.) and provides open source software that directly supports PKI for OpenSSH servers.

3. Uses

PKI has many uses, including providing public keys and bindings to user identities which are used for:

- Encryption and/or sender authentication of e-mail messages (e.g., using Open PGP or S/MIME).
- Encryption and/or authentication of documents (e.g., the XML Signature or XML Encryption standards if documents are encoded as XML).
- Authentication of users to applications (e.g., smart card logon, client authentication with SSL). There's experimental usage for digitally signed HTTP authentication in the Enigform and mod_openpgp projects.
- Bootstrapping[7] secure communication protocols, such as Internet Key Exchange (IKE) and SSL. In both of these, initial set-up of a secure channel (a "security association") uses asymmetric key—i.e., public key—methods, whereas actual communication uses faster symmetric key—i.e., secret key—methods.
- Mobile signatures are electronic signatures that are created using a mobile device and rely on signature or certification services in a location independent telecommunication environment.

New Words

create	*v.*	建立，创建
distribute	*v.*	分发，分配，发布，分布，分类，分区
revoke	*v.*	撤回，废除，宣告无效
ecommerce	*n.*	电子商务
confidential	*adj.*	秘密的，机密的
inadequate	*adj.*	不充分的，不适当的
rigorous	*adj.*	严格的，严厉的
proof	*n.*	证据，试验，考验
	adj.	不能透入的，防……的
	v.	检验，使不被穿透
validate	*v.*	确认，证实，验证
arrangement	*n.*	排列，安排
binding	*n.*	绑定，捆绑
establish	*v.*	建立，设立，确定
registration	*n.*	注册，报到，登记
issuance	*n.*	发行，发布
supervision	*n.*	监督，管理
assure	*v.*	保证，担保
valid	*adj.*	有效的，正当的，正确的
accept	*v.*	接受，认可，同意，承认
request	*v.* & *n.*	请求，要求

subordinate	*adj.* 次要的，从属的，下级的
	n. 下属
	v. 服从
identifiable	*adj.* 可以确认的
domain	*n.* 范围，领域
behalf	*n.* 为，利益
securely	*adv.* 安全地
reliably	*adv.* 可靠性
storage	*n.* 存储
distribution	*n.* 分配，分发
map	*v.* 映射
index	*n.* 索引，[数学]指数，指标
	v. 编入索引中，指出，做索引
delivery	*n.* 递送，交付，交货
certification	*n.* 证明
ecosystem	*n.* 生态系统
competitive	*adj.* 竞争的
dominate	*v.* 支配，占优势
temporary	*adj.* 暂时的，临时的
client	*n.* 顾客，客户，委托人
attestation	*n.* 证明
existence	*n.* 存在，实在
disjoint	*v.* 使不连接，(使)脱节，(使)解体
interoperate	*n.* 交互操作
introducer	*n.* 介绍人
grant	*v.* 同意，准予，承认(某事为真)
creator	*n.* 建立者，创建者
manual	*n.* 手册，指南
	adj. 手动的，手工的
accumulate	*v.* 积聚，堆积
designate	*v.* 指定，指派
gradually	*adv.* 逐渐地
expectation	*n.* 期待，预料，指望
decentralize	*n.* 分散
fault-tolerant	*adj.* 容错的
effort	*n.* 努力，成就
loop	*n.* 环，循环
blockchain	*n.* 区块链
unalterable	*adj.* 不能变更的

experimental	*adj.*	实验的
bootstrapping	*n.*	自引导，自举
channel	*n.*	信道，频道，通道
mobile	*adj.*	可移动的，机动的
device	*n.*	装置，设备
telecommunication	*n.*	电信，长途通信，无线电通信，电信学

Phrases

a set of	一组，一套
digital certificate	数字证书
Internet banking	网络银行
bind… with…	把……与……绑定
carry out	完成，实现，贯彻，执行
third-party validation authority	第三方验证机构
belong to	属于
separate from	分离，分开
trust relationship	信任关系，信托关系
account for	对……做出解释，导致，（比例）占
single sign-on	单点登录
put forth	发表，颁布，提出
authorization loop	授权环
suitable for…	适合……的
smart card	智能卡
mobile device	移动设备

Abbreviations

PKI (Public Key Infrastructure)	公共密钥基础设施
CA (Certificate Authority)	认证授权机构，证书颁发机构
RA (Registration Authority)	注册机构
VA (Validation Authority)	验证机构
WoT (Web of Trust)	信任网
SPKI (Simple Public Key Infrastructure)	简单公共密钥基础设施
TTP (Trusted Third Party)	可信第三方
MIME (Multipurpose Internet Mail Extension)	多用途网际邮件扩充协议
XML (eXtensible Markup Language)	可扩展标记语言
IKE (Internet Key Exchange)	因特网密钥交换

Notes

[1] In public key infrastructure, a Validation Authority (VA) is an entity that provides a service

used to verify the validity of a digital certificate per the mechanisms described in the X.509 standard and RFC 5280.

[2] In cryptography, a web of trust is a concept used in PGP, GnuPG, and other OpenPGP-compatible systems to establish the authenticity of the binding between a public key and its owner. Its decentralized trust model is an alternative to the centralized trust model of a Public Key Infrastructure (PKI), which relies exclusively on a certificate authority (or a hierarchy of such). As with computer networks, there are many independent webs of trust, and any user (through their identity certificate) can be a part of, and a link between, multiple webs.

[3] In cryptography, a Trusted Third Party (TTP) is an entity which facilitates interactions between two parties who both trust the third party; the third party reviews all critical transaction communications between the parties, based on the ease of creating fraudulent digital content. In TTP models, the relying parties use this trust to secure their own interactions.

[4] In cryptography, X.509 is a standard that defines the format of public key certificates. X.509 certificates are used in many Internet protocols, including TLS/SSL, which is the basis for HTTPS, the secure protocol for browsing the web. They're also used in offline applications, like electronic signatures. An X.509 certificate contains a public key and an identity (a hostname, or an organization, or an individual), and is either signed by a certificate authority or self-signed. When a certificate is signed by a certificate authority, or validated by another means, someone holding that certificate can rely on the public key it contains to establish secure communications with another party, or validate documents digitally signed by the corresponding private key.

[5] Philip R. "Phil" Zimmermann, Jr. (born February 12, 1954) is the creator of Pretty Good Privacy (PGP), the most widely used e-mail encryption software in the world. He is also known for his work in VoIP encryption protocols, notably ZRTP and Zfone.

[6] A blockchain—originally block chain—is a distributed database that is used to maintain a continuously growing list of records, called blocks. Each block contains a timestamp and a link to a previous block. A blockchain is typically managed by a peer-to-peer network collectively adhering to a protocol for validating new blocks. By design, blockchains are inherently resistant to modification of the data. Once recorded, the data in any given block cannot be altered retroactively without the alteration of all subsequent blocks and a collusion of the network majority. Functionally, a blockchain can serve as "an open, distributed ledger that can record transactions between two parties efficiently and in a verifiable and permanent way. The ledger itself can also be programmed to trigger transactions automatically".

[7] In general, bootstrapping usually refers to a self-starting process that is supposed to proceed without external input. In computer technology the term (usually shortened to booting) usually refers to the process of loading the basic software into the memory of a computer after power-on or general reset, especially the operating system which will then take care of loading other software as needed.

Exercises

[Ex. 1] Answer the following questions according to the text.

1. What does PKI stand for? What is it?
2. What is the purpose of a PKI?
3. What is an RA responsible for?
4. What does a PKI consist of?
5. What is the primary role of the CA?
6. What are the three certificate authorities which account for three-quarters of all issued certificates on public-facing web servers?
7. What is one of the benefits of the web of trust, such as in PGP?
8. Who first put forth the web of trust concept? And when and where?
9. What is an example of a blockchain-based cryptocurrency that supports the storage of different public key types (SSH, GPG, RFC 2230, etc.) and provides open source software that directly supports PKI for OpenSSH servers?
10. What are uses of PKI listed in the last part of the text?

[Ex. 2] Translate the following terms or phrases from English into Chinese and vice versa.

1. digital certificate	1.
2. mobile device	2.
3. single sign-on	3.
4. third-party validation authority	4.
5. trust relationship	5.
6. 绑定，捆绑	6.
7. 证明	7.
8. 信道，频道，通道	8.
9. 指定，指派	9.
10.装置，设备	10.

[Ex. 3] Translate the following passage into Chinese.

What Does Private Key Mean?

A private key is a tiny bit of code that is paired with a public key to set algorithms for text encryption and decryption. It is created as part of public key cryptography during asymmetric-key encryption and used to decrypt and transform a message to a readable format. Public and private keys are paired for secure communication, such as e-mail.

A private key is also known as a secret key.

A private key is shared only with the key's initiator, ensuring security. For example, A and B represent a message sender and message recipient, respectively. Each has its own pair of public and private keys. A, the message initiator or sender, sends a message to B. A's message is encrypted with B's public key, while B uses its private key to decrypt the message received from A.

A digital signature, or digital certificate, is used to ensure that A is the original message sender. To verify this, B uses the following steps:

- B uses A's public key to decrypt the digital signature, as A must previously use its private key to encrypt the digital signature or certificate.
- If readable, the digital signature is authenticated with a CA.

In short, sending encrypted messages requires that the sender use the recipient's public key and its own private key for encryption of the digital certificate. Thus, the recipient uses its own private key for message decryption, whereas the sender's public key is used for digital certificate decryption.

[Ex. 4] Fill in the blanks with the words given below.

registration	verifying	signature	procedures	bind
authenticate	sensitive	communities	distribution	trust

PKI

A Public Key Infrastructure (PKI) supports the distribution and identification of public encryption keys, enabling users and computers to both securely exchange data overnetworks such as the Internet and verify the identity of the other party.

Without PKI, ____1____ information can still been crypted (ensuring confidentiality) and exchanged, but there would be no assurance of the identity (authentication) of the other party. Any form of sensitive data exchanged over the Internet is reliant on PKI for security.

1. Elements of PKI

A typical PKI consists of hardware, software, policies and standards to manage the creation, administration, ____2____ and revocation of keys and digital certificates. Digital certificates are at the heart of PKI as they affirm the identity of the certificate subject and ____3____ that identity to the public key contained in the certificate.

A typical PKI includes the following key elements:

- A trusted party, called a CA, acts as the root of trust and provides services that authenticate the identity of individuals, computers and other entities.
- A ____4____ authority, often called a subordinate CA, certified by a root CA to issue certificates for specific uses permitted by the root.
- A certificate database, which stores certificate requests and issues and revokes certificates.
- A certificate store, which resides on a local computer as a place to store issued certificates and private keys.

A CA issues digital certificates to entities and individuals after ____5____ their identity. It signs these certificates using its private key; its public key is made available to all interested parties in a self-signed CA certificate. CAs use this trusted root certificate to create a "chain of trust"—many root certificates are embedded in Web browsers so they have built-in trust of those CAs. Web servers, e-mail clients, smartphones and many other types of hardware and software also support PKI and contain trusted root certificates from the major CAs.

Along with an entity's or individual's public key, digital certificates contain information about the algorithm used to create the ____6____, the person or entity identified, the digital signature of the

CA that verified the subject data and issued the certificate, the purpose of the public key encryption, signature and certificate signing, as well as a date range during which the certificate can be considered valid.

2. Problems with PKI

PKI provides a chain of ____7____, so that identities on a network can be verified. However, like any chain, a PKI is only as strong as its weakest link. There are various standards that cover aspects of PKI—such as the Internet X.509 Public Key Infrastructure Certificate Policy and Certification Practices Framework (RFC2527)—but there is no predominant governing body enforcing these standards. Although a CA is often referred to as a "trusted third party", shortcomings in the security ____8____ of various CAs in recent years has jeopardized trust in the entire PKI on which the Internet depends. If one CA is compromised, the security of the entire PKI is at risk. For example, in 2011, Web browser vendors were forced to blacklist all certificates issued by the Dutch CA DigiNotar after more than 500 fake certificates were discovered.

3. A Web of Trust

An alternative approach to using a CA to ____9____ public key information is a decentralized trust model called a "Web of trust", a concept used in PGP and other Open PGP-compatible systems. Instead of relying solely on a hierarchy of certificate authorities, certificates are signed by other users to endorse the association of that public key with the person or entity listed in the certificate. One problem with this method is a user has to trust all those in the key chain to be honest, so it's often best suited to small user ____10____. For example, an enterprise could use a web of trust for authenticating the identity of its internal, intranet and extranet users and devices. It could also act as its own CA, using software such as Microsoft Certificate Services to issue and revoke digital certificates.

Text B

The Basics of Website Security for E-commerce Retailers

For e-commerce retailers, website security is the cornerstone of a successful online business. Why? Because people only want to give their money and their business to companies and organizations that they can trust.

If a retailer has an insecure website, then, all other marketing and inbound efforts simply won't bring results. Here, we run through some basic security practices that all e-commerce retailers should employ to make sure that their website is a secure, successful online destination.

1. PCI Compliance

The PCI Security Standards Council is a global group formed to develop, enhance and maintain security standards for payment account security. Its founding members include American Express, Discover Financial Services, JCB International, MasterCard and Visa Inc.

Together, the members of this group has come up with a set of security requirements, known as the Payment Card Industry Data Security Standard (PCI DSS) that all merchants or organizations that process, store, or transmit credit card information must adhere to. There is good reason for this: These guidelines ensure that all stored credit card data is protected and that sensitive information is secure throughout the transaction process.

Many companies meet these guidelines through the use of tokenization.

Tokenization, when applied to data security, is the process of substituting a sensitive data element with a non-sensitive equivalent, referred to as a token, that has no extrinsic or exploitable meaning or value. When sensitive information, such as digits in a credit card number, is replaced by non-sensitive information, or tokens, it cannot be read. This is an effective means of encrypting data because it's extremely secure: The tokenized information can only be detokenized to redeem the sensitive data under strict security controls and the storage of tokens and payment card data must comply with PCI standards, including the use of strong cryptography.

Staying PCI compliant and ensuring that all stored credit card data is fully tokenized in this way greatly reduces the risk of this sensitive information being stolen and used. Keeping this data secure is extremely important for all online retailers. If a cardholder's data is stolen, their credit can be negatively affected and they could lose credibility, money, and even their business.

2. SSL[1] Certificate

The SSL Certificate, also mandatory per PCI, works to ensure that the sensitive information that is sent over the Internet is encrypted and secure as well. When retailers or site visitors send information or data over the Internet, it passes through multiple computers before reaching its destination server. At any point during this chain, it could get stolen if it is not encrypted with an SSL Certificate.

How does the certificate work? It essentially makes all sensitive information, which includes passwords, credit card information, and user names, unreadable for everyone except the destination server, thereby protecting all communication from eavesdropping and theft.

The SSL Certificate is particularly valuable for e-commerce retailers not just for security reasons but also to build trust with site visitors and prospective customers. Attaining an SSL certificate essentially verifies an entity's credentials, certifying that they are who they say they are and that their site is safe to visit.

Make sure to watch for changes in requirements, such as the recent change from SHA-1[2] encryption to SHA-2 encryption, to make sure your company stays compliant.

3. Use HTTPS

Hypertext Transfer Protocol with Secure Sockets Layer, or HTTPS, is a protocol to transfer data over the web that should be used instead of HTTP on all pages where data is created. Once again, the issue here is all about encryption. With HTTP, information is not encrypted—instead, it is sent as plain text, which means that anyone can intercept it and read what has been sent.

Furthermore, many customers know about this insecurity and tend to avoid e-commerce websites that use HTTP. This means that keeping HTTP could hurt a retailer's security and their business over time.

It's important to note, though, that HTTPS isn't necessary on every page of a website. Why? If retailers try to include it everywhere, it will slow their page load speed and likely hurt their business. HTTPS should just be used on pages that collect and store data so that customers can feel secure sending their information. That means skipping the homepage, about us page, etc.

4. DoS and DDOS protection

DoS (Denial of Service) and DDOS (Distributed Denial of Service) protection work to guard against denial of service and distributed denial of service attacks.

During both attacks, attackers attempt to block legitimate users from accessing information or services by flooding a network with requests, thereby overwhelming the bandwidth of the targeted system and preventing legitimate requests from coming through.

While both attacks work in the same way, the key difference is that a DoS attacker usually uses a single computer and Internet connection, while a DDoS attacker uses multiple connected devices, making the flood of information much larger and harder to deflect.

There are many ways to protect from DoS and DDoS attacks. The easiest and most expensive way is to buy more bandwidth. Assume that during these attacks, they're trying to flood your space. If you have a ton of space it will be more difficult for attackers to overwhelm you. However, this is a largely impractical solution, especially for DDoS attacks, since the attacks are just too large to overcome.

However, there are more inexpensive and effective other ways to mitigate attacks. Setting up effective, well-configured firewalls, for example, can prevent this attack traffic from reaching your computer.

5. Use a Firewall

As the name suggests, a firewall is a hardware or software system that essentially works as a wall or gateway between two or more networks. It permits authorized traffic and blocks unauthorized or potentially malicious traffic from accessing a network or system. Just like an actual wall.

A firewall essentially protects what is inside a network from the outside—a.k.a from other networks or from threats on the Internet like backdoor and DDoS attacks. Since e-commerce websites have a lot of inbound traffic, they need firewalls to protect themselves against malicious entry.

There are many different kinds of firewalls, but two very effective firewalls for online retailers are application gateways and proxy firewalls. Both function as intermediary programs between two or more networks, meaning that incoming traffic has no direct connection or access to a retailer's network.

6. Application Gateways

With an application gateway in place, there are two lines of communication: One between your computer and the proxy, the other between the proxy and the destination computer or network. It's essentially a checkpoint that all network information has to stop at. By serving as this middle point, the application gateways help hide and protect your network from others', and only let in traffic (or packets) that have been authorized.

7. Proxy Firewalls

Proxy firewalls are among the most secure. Why? Like the application gateway, the proxy serves as an intermediary connection. However, they take it one step further. Instead of your network connection going all the way through, a new network connection is started at the proxy firewall. This means that there is no direct connection between systems at all, which makes it even harder for attackers to discover your network and get in.

It is important to note that, it has to be properly configured to have a firewall to be effective. What does this mean? Well, firewalls don't automatically know which traffic is malicious, and they need to be programmed with this information. Make sure, then, that whoever sets up the firewall is properly configuring it so that all of the right information gets through.

By staying on top of all these security measures, online retailers can effectively build their customers' trust and their own company's reputability.

New Words

cornerstone	*n.*	基础
employ	*v.*	雇用，使用，利用
destination	*n.*	目的地
compliance	*n.*	兼容
organization	*n.*	组织，机构，团体
tokenization	*n.*	标记，标记化
equivalent	*n.*	等价物，相等物
	adj.	等价的，相等的，同意义的
extrinsic	*adj.*	外在的，外表的，外来的
tokenize	*v.*	标记，使令牌化
extremely	*adv.*	极端地，非常地
detokenize	*v.*	去令牌，脱令牌
redeem	*v.*	恢复
strict	*adj.*	严格的，严谨的；精确的
affected	*adj.*	受到影响的，受侵袭的
multiple	*adj.*	多样的，多重的
	n.	倍数，若干
	v.	成倍增加
chain	*n.*	链(条)，一连串，一系列
essentially	*adv.*	本质上，本来
unreadable	*adj.*	不可读的；难以理解的
valuable	*adj.*	有价值的
prospective	*adj.*	预期的

attain	*v.*	获得，达到
insecurity	*n.*	不安全
hurt	*v.*	危害，损害
overwhelming	*adj.*	压倒性的，无法抵抗的
deflect	*v.*	使中止，使放弃；偏离，偏转
overwhelm	*v.*	淹没，制服，压倒
solution	*n.*	解决办法，解决方案
intermediary	*adj.*	中间的，媒介的
	n.	中间物
application	*n.*	应用，应用程序，应用软件
checkpoint	*n.*	检查站，检查点
configure	*v.*	配置，设定
automatically	*adv.*	自动地，机械地
reputability	*adj.*	规范的

Phrases

run through	匆匆查阅；贯穿
Security Standards Council	安全标准委员会
come up with	提出，拿出
tend to	注意，趋向
application gateway	应用网关
proxy firewall	代理防火墙
get in	进入；到达
stay on top of	知道，掌握

Abbreviations

PCI (Payment Card Industry)	支付卡行业
DSS (Data Security Standard)	数据安全标准
SHA (Secure Hash Algorithm)	安全哈希算法
a.k.a. (also known as)	又名……，也叫作

Notes

[1] SSL was originally specified in the 1990s as a proprietary protocol that allowed Netscape browser clients using the Hypertext Transfer Protocol (HTTP) to communicate securely with Netscape web servers. SSL eventually came to be used to secure authentication and encryption for communication at the network transport layer.

[2] The Secure Hash Algorithms are a family of cryptographic hash functions published by the National Institute of Standards and Technology (NIST) as a U.S. Federal Information Processing Standard (FIPS).

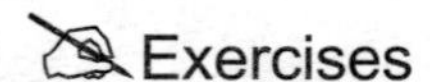Exercises

[Ex. 5] Answer the following questions according to the text.

1. Why is website security the cornerstone of a successful online business for e-commerce retailers?
2. What is the PCI Security Standards Council?
3. What does founding members of the PCI Security Standards Council include?
4. What is tokenization, when applied to data security?
5. Why is the SSL certificate particularly valuable for e-commerce retailers?
6. What does HTTPS stand for? What is it?
7. Why isn't HTTPS necessary on every page of a website?
8. What do DoS (Denial of Service) and DDOS (Distributed Denial of Service) protection do?
9. What is the difference between a DoS attacker and a DDoS attacker?
10. What are the two very effective firewalls for online retailers? What do they do?

Reading Material

Digital Signatures and Best Practices in IT Security

1. Digital Signatures

I have often had to answer the question "what is digital signature technology?" Most times, the question has been asked by regular people, smart but not highly technical and often the best way to explain[1] it is to draw parallels between electronic documents and paper documents. For example, when you go to your bank and want to withdraw some cash, you will sign a withdrawal slip[2] and sign on it then present it to the teller or cashier probably with your driver's license[3] or some other form of personal identification. The cashier will compare your signature with what the bank has in its records and if they match they will process your funds withdrawal. On the other hand, if some thief were to get hold of your identification papers and attempt to withdraw funds from your bank account, they would most likely fail because they would be unable to provide a signature that matches yours.

(1) What Is a Digital Signature

A digital signature is a mathematical scheme that is used to authenticate the sender of an electronic document. It ensures that the document is really from the sender and not from someone else while at the same time ensuring that the message that reaches the recipient is the same one sent without any alterations[4]. Digital signatures are very efficient in legally binding documents because

---

1 explain *v.* 解释，说明

2 withdrawal slip 取款单，提款单

3 Driver's license 驾驶执照

4 alteration *n.* 变更；改造

they are difficult to imitate[1] and can be time-stamped[2].

(2) How a Digital Signature Works

If you are sending a sensitive document, you would want the recipient of the document to know that it was from you and you would also want to ensure that the document gets to the recipient in the very same state you sent it in, without any alterations. The process of digitally signing your document would go something like this:

- First, you should copy the document and paste it into an e-mail note.
- Second, you use a special software to obtain a mathematical summary (commonly known as a message hash) of the contract.
- Thirdly, you will use a private key that you purchased from a trusted public-private key authority for encrypting the message hash.
- Lastly, you send your document with the message hash as your digital signature.

The digital signature can be used for signing any form of electronic document whether or not the message is encrypted. The digital signature is protected with a digital certificate that authenticates it. Your digital certificate will contain the certification—issuing authority's digital signature which makes it possible for anyone to verify that your certificate is real.

(3) Advantages of Digital Signatures

The following are the main benefits of using digital signatures:

- Speed: Businesses no longer have to wait for paper documents to be sent by courier[3]. Contracts are easily written, completed, and signed by all concerned parties in a little amount of time no matter how far the parties are geographically.
- Costs: Using postal or courier services for paper documents is much more expensive compared to using digital signatures on electronic documents.
- Security: The use of digital signatures and electronic documents reduces risks of documents being intercepted[4], read, destroyed[5], or altered[6] while in transit.
- Authenticity[7]: An electronic document signed with a digital signature can stand up in court just as well as any other signed paper document.
- Tracking[8]: A digitally signed document can easily be tracked and located in a short amount of time.
- Non-Repudiation: Signing an electronic document digitally identifies you as the signatory[9] and that cannot be later denied.

---

1 imitate *v.* 模仿，仿制，仿造
2 time-stamp 时间标记
3 courier *n.* 快递员，送快信的人
4 intercept *v.* 中途阻止，截取
5 destroy *v.* 破坏，毁坏
6 alter *v.* 改变
7 authenticity *n.* 确实性，真实性
8 track *v.* 追踪 *n.* 轨迹，跟踪
9 signatory *n.* 签名人，签字者

- Imposter prevention: No one else can forge your digital signature or submit[1] an electronic document falsely claiming it was signed by you.
- Time-Stamp: By time-stamping your digital signatures, you will clearly know when the document was signed.

(4) Disadvantages of Digital Signatures

Just like all other electronic products, digital signatures have some disadvantages that go with them. These include:

- Expiry: Digital signatures, like all technological products, are highly dependent on the technology it is based on. In this era of fast technological advancements, many of these tech products have a short shelf life.
- Certificates: In order to effectively use digital signatures, both senders and recipients may have to buy digital certificates at a cost from trusted certification authorities.
- Software: To work with digital certificates, senders and recipients have to buy verification software at a cost.
- Law: In some states and countries, laws regarding cyber and technology-based issues are weak or even non-existent. Trading in such jurisdictions[2] becomes very risky for those who use digitally signed electronic documents.
- Compatibility[3]: There are many different digital signature standards and most of them are incompatible[4] with each other and this complicates the sharing of digitally signed documents.

Most businesses today are embracing the idea of paperless offices[5]. To do that, they have identified what is a digital signature and the advantages of using them. They are now using digital signatures to authenticate important documents and make legally binding agreements[6].

2. Best Practices in IT Security

There are many best practices in IT security that are specific to certain industries or businesses, but some apply broadly.

(1) Balance Protection with Utility

Computers in an office could be completely protected if all the modems were torn out and everyone was kicked out of the room, but then they wouldn't be of use to anyone. This is why one of the biggest challenges in IT security is finding a balance between resource availability and the confidentiality and integrity of the resources.

Rather than trying to protect against all kinds of threats, most IT departments focus on

---

1 submit *v.* 提交，递交

2 jurisdiction *n.* 权限

3 compatibility *n.* 兼容性

4 incompatible *adj.* 不兼容的

5 paperless office 无纸化办公

6 agreement *n.* 协定，协议；同意；一致

insulating[1] the most vital systems first and then finding acceptable ways to protect the rest without making them useless. Some of the lower-priority systems may be candidates for automated analysis, so that the most important systems remain the focus.

(2) Split up[2] the Users and Resources

For an information security system to work, it must know who is allowed to see and do particular things. Someone in accounting, for example, doesn't need to see all the names in a client database, but he might need to see the figures coming out of[3] sales. This means that a system administrator needs to assign access by a person's job type, and may need to further refine those limits according to organizational separations[4]. This will ensure that the chief financial officer will ideally be able to access more data and resources than a junior accountant.

That said, rank doesn't mean full access. A company's CEO may need to see more data than other individuals, but he doesn't automatically[5] need full access to the system. This brings us to the next point.

(3) Assign Minimum Privileges

An individual should be assigned the minimum privileges needed to carry out his or her responsibilities[6]. If a person's responsibilities change, so will the privileges. Assigning minimum privileges reduces the chances that Joe from design will walk out the door with all the marketing data.

(4) Use Independent Defenses

This is a military principle[7] as much as an IT security one. Using one really good defense, such as authentication protocols, is only good until someone breaches it. When several independent defenses are employed, an attacker must use several different strategies to get through them. Introducing this type of complexity doesn't provide 100 percent protection against attacks, but it does reduce the chances of a successful attack.

(5) Plan for Failure

Planning for failure will help minimize its actual consequences[8] should it occur. Having backup systems in place beforehand allows the IT department to constantly monitor security measures and react quickly to a breach. If the breach is not serious, the business or organization can keep operating on backup while the problem is addressed. IT security is as much about limiting the damage from breaches as it is about preventing them.

(6) Record, Record, Record

Ideally, a security system will never be breached, but when a security breach does take place[9], the event should be recorded. In fact, IT staff often record as much as they can, even when a breach

---

1 insulate *v.* 使绝缘，隔离

2 split up 分离，分开

3 come out of … 由……产生，从……出来

4 separation *n.* 分离，分开

5 automatically *adv.* 自动地，机械地

6 responsibility *n.* 责任，职责

7 principle *n.* 法则，原则，原理

8 consequence *n.* 结果

9 take place 发生

isn't happening. Sometimes the causes of breaches aren't apparent after the fact, so it's important to have data to track backwards. Data from breaches will eventually help to improve the system and prevent future attacks, even if it doesn't initially make sense.

(7) Run Frequent Tests

Hackers are constantly improving their craft[1], which means information security must evolve to keep up[2]. IT professionals run tests, conduct risk assessments, reread the disaster recovery plan[3], plan[3], check the business continuity[4] plan in case of[5] attack, and then do it all over again.

参考译文

公钥基础设施

公钥基础设施（PKI）是创建、管理、分发、使用、存储和撤销数字证书和管理公钥加密所需的一系列角色、策略和过程。PKI 的目的是实现电子商务、网络银行和机密邮件等一系列网络活动相关信息的安全电子传输。在一些活动中，只简单地使用密码是不充分的认证方法，需要更严格的认证来确认参与通信各方的身份并验证正在传送的信息。这时就需要公钥基础设施。

在密码学中，PKI 将公钥与实体（如人员和组织）各自身份加以绑定。绑定通过在证书颁发机构（CA）注册和由其颁发证书来建立。根据绑定的保证级别，还可以通过自动化过程或在人力监督下进行（见图 8-1）。

确保有效和正确注册的 PKI 角色称为注册机构（RA）。RA 负责接受数字证书的请求并鉴别发出请求的实体。在 Microsoft PKI 中，注册机构通常称为从属 CA。

基于有关该实体的信息，每个 CA 域中的实体必须是唯一可识别的。第三方验证机构（VA）可以代表 CA 提供此实体信息。

1. 设计

公钥加密是一种加密技术，使实体能够在不安全的公共网络上安全通信，并通过数字签名可靠地验证实体的身份。

公共密钥基础设施（PKI）是用于创建、存储和分发数字证书的系统，用于验证特定公钥属于某个实体。PKI 创建将公钥映射到实体的数字证书，将这些证书安全地存储在中央存储库中，并在需要时撤销它们。

PKI 包括：

- 证书颁发机构（CA），存储、颁发和签署数字证书。
- 注册机构，核实请求要将数字证书存储在 CA 的实体的身份。
- 一个中央目录——即一个用于存储和索引密钥的安全位置。

---

1 craft　*n.* 工艺，手艺

2 keep up　维持，继续

3 disaster recovery plan　灾难恢复计划

4 continuity　*n.* 连续性，连贯性

5 in case of　假设，万一

• 证书管理系统，用于管理存储证书的访问权限或颁发证书。
• 证书政策。

2. 认证方式

一般来说，传统上有三种获得这种信任的方法：证书颁发机构（CA）、信任网（WoT）和简单公钥基础设施（SPKI）。

（1）证书颁发机构

CA 的主要作用是数字签名并发布与给定用户绑定的公钥。这通过使用 CA 自己的私钥来完成，所以对用户密钥的信任依赖于对 CA 密钥有效性的信任。当 CA 是与用户和系统分开的第三方时，则称为注册机构（RA），注册机构可能与 CA 分开也可能不分开。与用户绑定的密钥是根据绑定的保证级别通过软件或在人力监督下建立的。

可信第三方（TTP）也可以用于 CA。此外，PKI 本身也经常被用作 CA 实现的同义词。

1）发行人市场份额。在这种信任关系模型中，CA 是受可信第三方——证书的主体（所有者）和依赖证书的一方都信任它。

NetCraft 认为尽管全球生态系统是竞争的，但还是由一些主要的 CA 控制。三个认证机构（Symantec、Comodo 及 GoDaddy）占据公开网络服务器所用证书的四分之三。

2）临时证书和单点登录。此方法涉及在单点登录系统中充当离线证书颁发机构的服务器。单一登录服务器将向客户端系统发出数字证书，但永远不会将其存储。用户可以使用临时证书执行程序等。通常在使用基于 X.509 的多种证书中可以见到这种解决方案。

（2）信任网

解决公钥信息公开认证难题的另一种方法是使用信任网方案，它使用自签名证书和这些证书的第三方认证。“信任网络”的单数形式并不意味着存在一个单一的信任网络，或是共同的信任点，而是任何数量的潜在不相连的“信任网络”之一。实现该方法的示例是 PGP（Pretty Good Privacy）和 GnuPG（实现 OpenPGP，PGP 的标准化规范）。由于 PGP 和实现允许使用电子邮件数字签名来公开公钥信息，因此实现自己的信任网络相对容易。

信任网络的好处之一，如在 PGP 中，是它可以与一个区域（如公司内部的 CA）中各方完全信任的 PKI CA 进行互操作，该 PKI CA 愿意为证书担保，充当值得信赖的介绍人。如果对“信任网络”完全信任，那么根据信任网络的性质，信任一个证书就是对该网络中所有证书的信任。

信任网络概念由 PGP 创始人 Phil Zimmermann 于 1992 年在 PGP 2.0 版手册中提出。随着时间的推移，你会积累一些来自其他人的密钥，你可能希望将其指定为可信的介绍人。每个人都会选择自己值得信赖的介绍人。每个人都将逐渐积累和分发来自其他人的认证签名，期望任何收到签名的人将至少信任一两个签名。对所有公钥而言，这将导致出现分散的容错信任网络。

（3）简单公钥基础设施

另一种不涉及公共密钥信息公开认证的替代方案是简单公钥基础设施（SPKI），它从三个独立的成果中诞生，以克服 X.509 和 PGP 的信任网络的复杂性。SPKI 不会将用户与人员相关联，因为它信赖的是密钥而不是人。SPKI 不使用任何信任概念，因为验证者也是发行人。这在 SPKI 术语中被称为“授权循环”，其中授权是其设计的组成部分。

（4）基于区块链的 PKI

PKI 的新兴方法是使用通常与现代密码学相关的区块链技术。由于区块链技术旨在提供分布式和不可更改的信息分类账，因此它具有高度适用于公钥存储和管理的品质。Emercoin 是基于区块链的密码学实例，支持不同公钥类型（SSH、GPG 及 RFC 2230 等）的存储，并提供直接支持 OpenSSH 服务器 PKI 的开源软件。

3. 用途

PKI 有许多用途，包括提供公共密钥和绑定用户身份，这可用于：

- 电子邮件消息的加密和/或发件人身份验证（例如，使用 OpenPGP 或 S/MIME）。
- 文档的加密和/或验证（例如，如果文档被编码为 XML，则为 XML 签名或 XML 加密标准）。
- 用户对应用程序的身份验证（例如，智能卡登录，使用 SSL 的客户端身份验证）。在 Enigform 和 mod_openpgp 项目中有数字签名的 HTTP 身份验证的实验用法。
- 自引导安全通信协议，如因特网密钥交换（IKE）和 SSL。在这两者中，安全通道的初始设置（“安全关联”）使用非对称密钥——即公共密钥——方法，而实际通信使用更快的对称密钥——即秘密密钥——方法。
- 移动签名是使用移动设备创建的电子签名，并且在独立电信环境中依靠签名或认证服务。

Unit 9

Text A

Digital Authentication

The potential for fraud is always a risk that cannot be ignored when it comes to conducting transactions. In person, an individual could present forged or altered documents that attest to an identity that does not belong to him or her.

Online, an individual could also misrepresent his or her identity in a similar manner using someone else's credentials without their permission.

1. What is authentication?

The term authentication refers to an electronic process that allows for the electronic identification of a natural or legal person. Additionally, authentication may also confirm the origin and integrity of data in electronic form, such as the issuance of a digital certificate to attest to the authenticity of a website. The overall purpose of authentication is to reduce the potential for fraud, especially in the event of an individual purposely misrepresenting their identity or through the unauthorized use of another person's credentials.

The term digital authentication or electronic authentication (e-authentication[1]) refers to the process where the confidence in user identities is established and presented electronically to an information system.

The digital authentication process presents a technical challenge due to the necessity of authenticating individual people or entities remotely over a network. Its level of security depends on the applied type of authentication, the authentication factors used, as well as the process of authentication applied.

2. Authentication Factors

When authenticating an online user, there are three factor categories that may be used to assure that the user is who he or she makes a claim to be. These factor categories are:

- Knowledge factors—these would include a user's password, passphrase, personal identification number (pin) or a challenge response where the user would be required to answer a preselected security question.
- Ownership factors—these would include something that the user has possession of, such as a bank card, a hardware or software, One Time Password (OTP) token or a cell phone.

- Inherence factors—these factors relate to something that a user is or does and includes biometric identifiers such as facial, fingerprint or retinal pattern recognition along with other personal trait identifiers.

3. Types of Authentication

The subsequent categorization lists the most frequently used types of online user authentication sorted based on increasing levels of security:

1) Single-factor authentication—only one component out of one of the 3 factor categories is used to authenticate a person's identity. Experience shows that one single factor does not provide sufficient protection against malicious intrusion and misuse. Therefore, as soon as financially or personally relevant transactions are involved, a higher level of security is preferable.

2) Two-factor authentication—often referred to as 2FA, the user's identity is confirmed by using a combination of two independent components from two different factor categories. For example, where a user has logged on to their online bank account, with their username and password, and wishes to complete an online transaction, he or she would need to enter an authentication factor in addition to the knowledge factor (username and password) that was used to log on. The additional factor must also be from a different factor category than the username and password. An online banking user would typically use an authentication mechanism from the ownership category such as an OTP device or mobile phone to receive an OTP in a text message. OTPs are dynamic passwords which can only be used once and thereby provide a strong level of protection against a range of attacks.

3) Multi-factor authentication is similar to 2FA, but it can combine more than 2 authentication factors for enhanced security, whereas 2FA only uses two different factors.

4) Strong authentication—this type is often used as synonym for multi-factor authentication or 2FA. However, unlike multi-factor authentication and 2FA, strong authentication mandatorily requires non replicable factors or the use of digital certificates to provide a higher level of authentication for users. If those criteria are fulfilled, multi-factor authentication and 2FA are able to provide strong authentication.

The European Central Bank (ECB) defines strong customer authentication as a procedure based on two or more of the three authentication factors: ①knowledge, ②ownership, and③inherence. According to the ECB definition, the individual elements that are chosen for strong authentication factors must be mutually independent and at least one must be non-reusable and non-replicable (except for inherence), and not capable of being surreptitiously stolen via the Internet. What does that mean? Independence is relatively easy to grasp—if one component is related to the other, hacking one means all components are tampered with. Non-replicability implies the aspect of time or usage. Used once it cannot be used again. An example is a one-time-password which is only valid during a short period of time (e.g., 30 seconds).

In the United States, the National Information Assurance (IA) Glossary produced by the Committee on National Security Systems defines strong authentication similarly, requiring multiple factors for authentication and advanced technology, such as dynamic passwords or digital certificates to verify an entity's identity.

4. Enrolment and Authentication Process

The American National Institute of Standards and Technology (NIST) has outlined a quite generic digital authentication model, which can be used as a basic explanation model for the authentication process, regardless of the geographical region or area of jurisdiction.

In the NIST model, an individual (applicant) applies to a credential service provider (CSP)[2] and thus initiates the enrollment process. Once the CSP has successfully proven the applicant's identity, he or she becomes a "subscriber", and an authenticator (e.g., token) as well as a corresponding credential, such as a username, are established between the CSP and the applicant (now owning the role of the subscriber).

The CSP has the task of maintaining the credential including its status and all enrollment data over the whole lifetime of the credential. The subscriber needs to maintain the authenticator. This first part of the NIST reference model is applied in any enrollment process, where subsequent authentication is required, e.g., when a bank account is created or when a person signs up in an e-government process.

Once the applicant has become a "subscriber", he or she can perform online transactions within an authenticated session, conducted with a relying party. In such a transaction, the person holds the role of a claimant, proving to a verifier the possession of one or more authenticators. Verifier and relying party might be the same or alternatively two independent entities. If verifier and relying party are separate, the verifier has to provide assertion about the subscriber to the relying party. Subsequent to this assertion, the relying party may then initiate the transaction process.

In the following, a typical payment sequence illustrates this reference process:

The account owner (subscriber) wants to initiate a transaction. He or she first needs to prove through one or more authenticators that he or she, who claims to be the account owner (claimant) actually is the person he claims to be (subscriber). The validation is done by a "verifier" who verifies the authenticators at the "credential service provider" and after validation gives authentication assertion to the transaction department of the bank (relying party). In many banks the entities "verifier" and "credential service provider" may well be entities within the bank.

E-government processes are conducted in a similar way (e.g. the application for a passport). But in many cases, the "credential service providers" are external entities.

5. Conclusion

The burden of fraudulent transactions falls upon individuals, businesses and financial institutions with losses that result in costs passed down to consumers along with costs related to identity theft. Authentication is one means to protect online transactions, along with their sender or recipient from falling victim to fraud.

Other concepts such as digital signing can also leverage (strong) authentication to achieve better security, data integrity and non-repudiation for digital transactions.

New Words

transaction	*n.*	交易；事务
forge	*v.*	伪造
online	*n.*	联机，在线式
misrepresent	*v.*	误传，不如实地叙述(或说明)，说假话
credential	*n.*	信任状
permission	*n.*	许可，允许
identification	*n.*	辨认，鉴定，证明
origin	*n.*	起源，由来
purposely	*adv.*	故意地
necessity	*n.*	必要性，需要
factor	*n.*	因素，要素
preselect	*v.*	预先选择，预先选定
possession	*n.*	拥有，占有，所有
inherence	*n.*	内在，固有
facial	*adj.*	面部的
retinal	*adj.*	视网膜的
trait	*n.*	显著的特点，特性
subsequent	*adj.*	后来的，并发的
preferable	*adj.*	更可取的，更好的，更优越的
enter	*v.* & *n.*	输入
mechanism	*n.*	机制
dynamic	*adj.*	动态的
enhanced	*adj.*	增强的，提高的，放大的
mandatorily	*adv.*	命令地，强制地
replicable	*adj.*	可复制的，能复现的
fulfill	*v.*	履行，实现，完成
define	*v.*	定义，详细说明
mutually	*adv.*	互相地
non-reusable	*adj.*	不可以再度使用的，不可重复使用的
surreptitiously	*adv.*	秘密地
grasp	*v.* & *n.*	抓住，抓紧；掌握，领会
non-replicability	*n.*	不可复制性，不可复现
imply	*v.*	暗示，意味
enrolment	*n.*	登记
outline	*v.*	描画轮廓，略述
	n.	大纲，轮廓，略图；外形；要点，概要
generic	*adj.*	一般的，普通的，非特殊的

jurisdiction	*n.*	权限
applicant	*n.*	申请者，请求者
subscriber	*n.*	签署者，订户
status	*n.*	身份，地位，情形，状况
e-government	*n.*	电子政务
session	*n.*	会话期
verifier	*n.*	证实者，核验者，证明者
payment	*n.*	付款，支付
illustrate	*v.*	举例说明，图解，阐明
claimant	*n.*	（根据权利）提出要求者
validation	*n.*	确认
burden	*n.*	担子，负担
	v.	负担
leverage	*n.*	杠杆作用

Phrases

attest to	证实，证明
in a manner	在某种意义上
digital authentication	数字证书
make a claim to…	认为……是属于自己的
personal identification number (pin)	个人身份号码
biometric identifier	生物标志
pattern recognition	模式识别
along with…	连同……一起，随同……一起
as soon as…	一……就
be confirmed by…	由……确认，用……证实
independent component	独立部分
log on	登录
bank account	银行账户
online transaction	在线交易，在线业务
in addition to…	除……之外
multi-factor authentication	多因素认证，多重认证
strong authentication	强认证，强身份验证，强验证
individual element	单个元素
except for…	除……以外
tamper with	损害，篡改
National Information Assurance (IA) Glossary	国家信息保障（IA）词汇
aspect of time	时间方面
reference model	参考模型

fall upon 开始行动，进攻

Abbreviations

OTP (One Time Password) 一次性密码
2FA (Two-Factor Authentication) 双因素认证，双重认证
ECB (European Central Bank) 欧洲中央银行
IA (Information Assurance) 信息保障
NIST (American National Institute of Standards and Technology) 美国国家标准与技术研究所
CSP (Credential Service Provider) 证书服务提供商

Notes

[1] Electronic authentication is the process of establishing confidence in user identities electronically presented to an information system. Digital authentication or e-authentication may be used synonymously when referring to the authentication process that confirms or certifies a person's identity and works. When used in conjunction with an electronic signature, it can provide evidence whether data received has been tampered with after being signed by its original sender. In a time where fraud and identity theft has become rampant, electronic authentication can be a more secure method of verifying that a person is who they say they are when performing transactions online.

[2] A Credential Service Provider (CSP) is a trusted entity that issues security tokens or electronic credentials to subscribers. A CSP forms part of anauthentication system, most typically identified as a separate entity in a Federated authentication system. A CSP may be an independent third party, or may issue credentials for its own use. The term CSP is used frequently in the context of the US government's eGov and e-authentication initiatives. An example of a CSP would be an online site whose primary purpose may be, for example, Internet banking—but whose users may be subsequently authenticated to other sites, applications or services without further action on their part.

Exercises

[Ex. 1] Answer the following questions according to the text.

1. What does the term authentication refer to?

2. What is the overall purpose of authentication?

3. What does the term digital authentication or electronic authentication (e-authentication) refer to?

4. What are three factor categories that may be used to assure that the user is who he or she makes a claim to be when authenticating an online user?

5. What are the most frequently used types of online user authentication sorted based on increasing levels of security?

6. What does strong authentication mandatorily require to provide a higher level of authentication for users?

7. According to the ECB's definition on strong customer authentication, what must the

individual elements that are chosen for strong authentication factors be?

8. What does an individual do to initiate the enrollment process in the NIST model?

9. What task does the CSP have?

10. Once the applicant has become a "subscriber", what can he or she do?

[Ex. 2] Translate the following terms or phrases from English into Chinese and vice versa.

1. reference model	1.
2. pattern recognition	2.
3. multi-factor authentication	3.
4. digital authentication	4.
5. biometric identifier	5.
6. 信任状	6.
7. 因素，要素	7.
8. 辨认，鉴定，证明	8.
9. 机制	9.
10.不可复制性，不可复现	10.

[Ex. 3] Translate the following passage into Chinese.

If you are serious about computer/network security, then you must have a solid understanding of authentication methods.

Computer/network security hinges on two very simple goals:

1) Keeping unauthorized persons from gaining access to resources.

2) Ensuring that authorized persons can access the resources they need.

There are a number of components involved in accomplishing these objectives. One way is to assign access permissions to resources that specify which users can or cannot access those resources and under what circumstances. (For example, you may want a specific user or group of users to have access when logged on from a computer that is physically on-site but not from a remote dial-up connection.)

Access permissions, however, work only if you are able to verify the identity of the user who is attempting to access the resources. That's where authentication comes in. In this passage, we will look at the role played by authentication in a network security plan, popular types of authentication, how authentication works, and the most commonly used authentication methods and protocols.

[Ex. 4] Fill in the blanks with the words given below.

infrastructure	session	status	contain	ownership
governments	minimizing	root	accepted	compatibility

What are Certificate Authorities & Trust Hierarchies?

An SSL Certificate is a popular type of Digital Certificate that binds the ownership details of a web server (and website) to cryptographic keys. These keys are used in the SSL/TLS protocol to activate a secure ___1___ between a browser and the web server hosting the SSL Certificate. In order for a browser to trust an SSL Certificate, and establish an SSL/TLS session without security warnings, the SSL Certificate must ___2___ the domain name of website using it, be issued by a trusted CA, and not have expired.

According to an analyst site, in August 2015 there are almost 5.5m SSL Certificates in use for public facing websites. This makes SSL one of the most prevalent security technologies in use today.

1. With all these SSL Certificates in use, who decides a CA can be trusted?

Browsers, operating systems, and mobile devices operate authorized CA membership programs where a CA must meet detailed criteria to be ___3___ as a member. Once accepted the CA can issue SSL Certificates that are transparently trusted by browsers, and subsequently, people and devices relying on the certificates. There are a relatively small number of authorized CAs, from private companies to ___4___, and typically the longer the CA has been operational, the more browsers and devices will trust the certificates the CA issues. For certificates to be transparently trusted, they must have significant backward ___5___ with older browsers and especially older mobile devices—this is known as ubiquity and is one of the most important features a CA can offer its customers.

Prior to issuing a Digital Certificate, the CA will conduct a number of checks into the identity of the applicant. The checks relate to the class and type of certificate being applied for. For example, a domain validated SSL Certificate will have verified the ___6___ of the domain to be included within the Certificate, whereas an Extended Validation SSL will include additional information on the company, verified by the CA through many company checks.

2. Trust Hierarchies

Browsers and devices trust a CA by accepting the Root Certificate into its root store—essentially a database of approved CAs that come pre-installed with the browser or device. Windows operates a root store, as does Apple, Mozilla (for its Firefox browser) and typically each mobile carrier also operates its own ___7___ store.

CAs use these pre-installed Root Certificates to issue Intermediate Root Certificates (ICA) and end entity Digital Certificates, such as SSL Certificates. The CA receives certificate requests, validates the applications, issues the certificates, and publishes the ongoing validity ___8___ of issued certificates so anyone relying on the certificate has a good idea that the certificate is still valid.

The GlobalSign Extended Validation CA - G2 is shown in this example as the ICA—it's trust is inherited from the publicly trusted GlobalSign root (top of the hierarchy). This ICA is able to issue publicly trusted end entity certificates. In this example, the ICA issued an Extended Validation Certificate to www.globalsign.com.

CAs should not issue Digital Certificates directly from the root distributed to the carriers, but instead via one or more of their ICAs. This is because a CA should follow best security practices by ___9___ the potential exposure of a Root CA to attackers. GlobalSign is one of the few CAs to have always (since 1996) utilized ICAs.

3. What goes into running a CA?

As a trust anchor for the Internet, CAs have significant responsibility. As such running a CA within the auditable requirements is a complex task. A CA's ___10___ consists of considerable operational elements, hardware, software, policy frameworks and practice statements, auditing, security infrastructure and personnel. Collectively the elements are referred to as a trusted PKI.

Text B

Electronic Authentication

Electronic authentication, also referred to as e-authentication is the process of establishing confidence in user identities electronically presented to an information system. In online environments, the username identifies the user, while the password authenticates that the user is whom he or she claims to be. There are various ways to increase transactional security through e-authentication, often known as Multi-Factor Authentication (MFA)[1]. Methods can be a security token, captcha[2], or a challenge question. Users can only continue with their activities such as completing a transaction or accessing more information from a database, after their identity has been verified through e-authentication.

Accounts with a highly complicated registration or with more user restrictions are usually secured with more advanced e-authentication procedures. E-authentication also presents a technical challenge when this process involves the remote authentication of individual people over a network, for the purpose of electronic government and commerce.

1. Overview

In the conceptual e-authentication model, a claimant in an authentication protocol is a subscriber to some Credential Service Provider (CSP)[3]. At some point, an applicant registers with a Registration Authority (RA)[4], which verifies the identity of the applicant, typically through the presentation of paper credentials and by records in databases. This process is called identity proofing. The RA, in turn, vouches for the identity of the applicant (and possibly other verified attributes) to a CSP. The applicant then becomes a subscriber of the CSP. The CSP establishes a mechanism to uniquely identify each subscriber and the associated tokens and credentials issued to that subscriber. There is always a relationship between the RA and CSP. In the simplest and perhaps the most common case, the RA/CSP are separate functions of the same entity. However, an RA might be part of a company or organization that registers subscribers with an independent CSP, or several different CSPs. Therefore, a CSP may have an integral RA, or it may have relationships with multiple independent RAs, and an RA may have relationships with different CSPs as well.

2. E-authentication Methods

Authentication systems are often categorized by the number of factors that they incorporate. The three factors often considered as the cornerstone of authentication are: Something you know (for example, a password); something you have (for example, an ID badge or a cryptographic key); something you are (for example, a voice print, thumb print or other biometric).

multi-factor authentication is generally more secure than single-factor authentication. But, some multi-factor authentication approaches are still vulnerable to cases like man-in-the-middle attacks and Trojan attacks. Common methods used in authentication systems are summarized below.

(1) Token

Tokens generically are something the claimant possesses and controls that may be used to authenticate the claimant's identity (see Figure 9-1). In e-authentication, the claimant authenticates to a system or application over a network. Therefore, a token used for e-authentication is a secret and the token must be protected. The token may, for example, be a cryptographic key, that is protected by encrypting it under a password. An impostor must steal the encrypted key and learn the password to use the token.

Figure 9-1 A sample of token

(2) Passwords and PIN-Based Authentication

Passwords and PINs are categorized as "something you know" method. A combination of numbers, symbols, and mixed cases are considered to be stronger than all-letter password. Also, the adoption of Transport Layer Security (TLS) or Secure Socket Layer (SSL) features during the information transmission process will as well create an encrypted channel for data exchange and to further protect information delivered. Currently, most security attacks target on password-based authentication systems.

(3) Public-Key Authentication

This type of authentication has two parts. One is a public key, the other is a private key. A public key is issued by a Certification Authority and is available to any user or server. A private key is known by the user only.

(4) Symmetric-Key Authentication

The user shares a unique key with an authentication server. When the user sends a randomly generated message (the challenge) encrypted by the secret key to the authentication server, if the message can be matched by the server using its shared secret key, the user is authenticated. When implemented together with the password authentication, this method also provides a possible solution for two-factor authentication systems.

(5) SMS-Based Authentication

The user receives password by reading the message in the cell phone, and types back the password to complete the authentication. Short Message Service (SMS) is very effective when cell phones are commonly adopted. SMS is also suitable against Man-in-the-Middle (MITM) attacks, since the use of SMS does not involve the Internet.

(6) Biometric Authentication

Biometric authentication authenticates individuals by identifying their physiological or behavioral characteristic. The physiological features include but are not limited to fingerprint, hand geometry, retina scan, iris scan, signature dynamics, keyboard dynamics, voice print, and facial scan. The most widely used biometric authentication nowadays is the fingerprint method (see Figure 9-2).

Figure 9-2 Biometric authentication

(7) Digital Identity Authentication

Digital identity authentication refers to the combined use of device, behavior, location and other data, including e-mail address, account and credit card information to authenticate online users in real time.

3. Electronic Credentials

Paper credentials are documents that attest to the identity or other attributes of an individual or entity called the subject of the credentials. Some common paper credentials include passports, birth certificates, driver's licenses, and employee identity cards. The credentials themselves are authenticated in a variety of ways: Traditionally perhaps by a signature or a seal, special papers and inks, high quality engraving, and today by more complex mechanisms, such as holograms, that make the credentials recognizable and difficult to copy or forge. In some cases, simple possession of the credentials is sufficient to establish that the physical holder of the credentials is indeed the subject of the credentials. More commonly, the credentials contain biometric information such as the subject's description, a picture of the subject or the handwritten signature of the subject that can be used to authenticate that the holder of the credentials is indeed the subject of the credentials. When these paper credentials are presented in-person, authentication biometrics contained in those credentials can be checked to confirm that the physical holder of the credential is the subject.

Electronic identity credentials bind a name and perhaps other attributes to a token. There are a variety of electronic credential types in use today, and new types of credentials are constantly being created. At a minimum, credentials include identifying information that permits recovery of the records of the registration associated with the credentials and a name that is associated with the subscriber.

4. Verifiers

In any authenticated on-line transaction, the verifier is the party that verifies that the claimant has possession and control of the token that verifies his or her identity. A claimant authenticates his or her identity to a verifier by the use of a token and an authentication protocol. This is called Proof of Possession (PoP). Many PoP protocols are designed so that a verifier, with no knowledge of the token before the authentication protocol run, learns nothing about the token from the run. The verifier and CSP may be the same entity, the verifier and relying party[5] may be the same entity or they may all three be separate entities. It is undesirable for verifiers to learn shared secrets unless they are a part of the same entity as the CSP that registered the tokens. Where the verifier and the relying party are separate entities, the verifier must convey the result of the authentication protocol to the relying party. The object created by the verifier to convey this result is called an assertion.

5. Risk Assessment

When developing electronic systems, there are some industry standards requiring United States agencies to ensure the transactions provide an appropriate level of assurance. Generally, servers adopt the US' Office of Management and Budget's (OMB's) E-Authentication Guidance for Federal Agencies (M-04-04) as a guideline, which is published to help federal agencies provide secure electronic services that protect individual privacy. It asks agencies to check whether their transactions require e-authentication, and determine a proper level of assurance.

It established four levels of assurance:

Assurance Level 1: Little or no confidence in the asserted identity's validity.

Assurance Level 2: Some confidence in the asserted identity's validity.

Assurance Level 3: High confidence in the asserted identity's validity.

Assurance Level 4: Very high confidence in the asserted identity's validity.

(1) Determining Assurance Levels

The OMB proposes a five-step process to determine the appropriate assurance level for their applications:

- Conduct a risk assessment, which measures possible negative impacts.
- Compare with the four assurance levels and decide which one suits this case.
- Select technology according to the technical guidance issued by NIST.
- Confirm the selected authentication process satisfies requirements.
- Reassess the system regularly and adjust it with changes.

The required level of authentication assurance are assessed through the factors below:

- Inconvenience, distress, or damage to standing or reputation;
- Financial loss or agency liability;
- Harm to agency programs or public interests;
- Unauthorized release of sensitive information;
- Personal safety; and/or civil or criminal violations.

(2) Determining Technical Requirements

National Institute of Standards and Technology (NIST) guidance defines technical requirements for each of the four levels of assurance in the following areas:

- Tokens are used for proving identity. Passwords and symmetric cryptographic keys are private information that the verifier needs to protect. Asymmetric cryptographic keys have a private key (which only the subscriber knows) and a related public key.
- Identity proofing, registration, and the delivery of credentials that bind an identity to a token. This process can involve a far distance operation.
- Credentials, tokens, and authentication protocols can also be combined together to identify that a claimant is in fact the claimed subscriber.
- An assertion mechanism that involves either a digital signature of the claimant or is acquired directly by a trusted third party through a secure authentication protocol.

New Words

claim	*v.*	声称，(根据权利)要求
	n.	主张，(根据权利提出)要求
multi-factor	*adj.*	多种因素(或成分、原因等)的，多因子的
conceptual	*adj.*	概念上的
proofing	*n.*	证明，试验(法)，验算
impostor	*n.*	冒名顶替者
randomly	*adv.*	随机地
suitable	*adj.*	适当的，相配的
involve	*v.*	包括，包含，参与
physiological	*adj.*	生理学的，生理学上的
behavioral	*adj.*	行为的
engraving	*n.*	雕版，雕版图
hologram	*n.*	全息摄影，全息图
convey	*v.*	传达，转让
validity	*n.*	有效性，合法性，正确性
satisfy	*v.*	满足，确保
reassess	*v.*	再估价，再评价

Phrases

electronic authentication	电子认证
subscribe to	用户，同意，预订，订阅
identity proofing	身份证明
vouch for	担保，保证
voice print	声纹
thumb print	指纹
Man-in-the-Middle Attack(MITM)	中间人攻击
data exchange	数据交换

hand geometry	手形，掌形
retina scan	视网膜扫描
facial scan	面部扫描
handwritten signature	亲笔签字
separate entity	单独实体，独立实体
risk assessment	风险评估
individual privacy	个人隐私
negative impact	负面影响
public interest	公共利益

Abbreviations

MFA (Multi-Factor Authentication) 多因素认证，多因素身份验证

CAPTCHA (Completely Automated Public Turing Test to Tell Computers and Humans Apart) 全自动区分计算机和人类的图灵测试

TLS (Transport Layer Security) 传输层安全

PoP (Proof of Possession) 所有权证明

OMB (Office of Management and Budget) 管理和预算办公室

Notes

[1] Multi-Factor Authentication (MFA) combines two or more independent credentials: What the user knows (password), what the user has (security token) and what the user is (biometric verification). The goal of MFA is to create a layered defense and make it more difficult for an unauthorized person to access a target such as a physical location, computing device, network or database. If one factor is compromised or broken, the attacker still has at least one more barrier to breach before successfully breaking into the target.

[2] A captcha is a type of challenge-response test used in computing to determine whether or not the user is human.

[3] A Credential Service Provider (CSP) is a trusted entity that issues security tokens or electronic credentials to subscribers. A CSP forms part of an authentication system, most typically identified as a separate entity in a federated authentication system. A CSP may be an independent third party, or may issue credentials for its own use.

[4] A Registration Authority (RA) is an authority in a network that verifies user requests for a digital certificate and tells the Certificate Authority (CA) to issue it. RAs are part of a Public Key Infrastructure (PKI), a networked system that enables companies and users to exchange information and money safely and securely. The digital certificate contains a public key that is used to encrypt and decrypt messages and digital signatures.

[5] A Relying Party (RP) is a computer term used to refer to a server providing access to a secure software application.

Exercises

[Ex. 5] Fill in the following blanks according to the text.

1. Electronic authentication, also referred to as ________________ is the process of establishing confidence in ________________ electronically presented to an information system.

2. In the conceptual e-authentication model, a claimant in an authentication protocol is ________________ to some ________________.

3. The three factors often considered as the cornerstone of authentication are: ____________________________________, ____________________________________, and ________________________________.

4. Multi-factor authentication is generally ______________ than single-factor authentication. But, some multi-factor authentication approaches are still vulnerable to cases like ________________ and ________________.

5. Tokens generically are________________and controls that may be used to ______________________.

6. Passwords and PINs are categorized as __________________. A combination of numbers, symbols, and mixed cases are considered _________________________.

7. Public-key authentication has two parts. One is ______________, the other is ______________. The former is issued by a ______________ and is available to any user or server. The latter is known by ______________.

8. Short Message Service (SMS) is very effective when __________________________. SMS is also suitable against Man-in-the-Middle (MITM) attacks, since the use of SMS ___________________.

9. Digital identity authentication refers to the combined use of device, ___________, ______________ and other data, including ____________, account and __________________to authenticate online users in real time.

10. The OMB proposes a five-step process to determine the appropriate assurance level for their applications:

- ____.
- ____.
- ____.
- ____.
- ____.

Reading Material

Data Backup

In information technology, a backup, or the process of backing up, refers to the copying and archiving[1] of computer data so it may be used to restore the original after a data loss event. The verb form is to back up in two words, whereas the noun is backup.

Backups have two distinct purposes. The primary purpose is to recover[2] data after its loss, be it by data deletion or corruption. The secondary purpose of backups is to recover data from an earlier time, according to a user-defined data retention[3] policy, typically configured within a backup application for how long copies of data are required. Though backups represent a simple form of disaster recovery, and should be part of any disaster[4] recovery plan, backups by themselves should not be considered a complete disaster recovery plan. One reason for this is that not all backup systems are able to reconstitute[5] a computer system or other complex configuration such as a computer cluster, active directory server, or database server by simply restoring data from a backup.

Since a backup system contains at least one copy of all data considered worth saving, the data storage requirements can be significant. Organizing this storage space and managing the backup process can be a complicated undertaking. A data repository model[6] may be used to provide structure to the storage. Nowadays, there are many different types of data storage devices that are useful for making backups. There are also many different ways in which these devices can be arranged to provide geographic redundancy[7], data security, and portability[8].

Before data are sent to their storage locations, they are selected, extracted, and manipulated. Many different techniques have been developed to optimize the backup procedure. These include optimization for dealing with open files and live data sources as well as compression[9], encryption, and deduplication[10], among others. Every backup scheme should include dry runs[11] that validate the reliability of the data being backed up. It is important to recognize the limitations and human factors involved in any backup scheme.

1. Selection and Extraction[12] of Data

A successful backup job starts with selecting and extracting coherent[13] units of data. Most data on modern computer systems is stored in discrete[14] units, known as files[15]. These files are organized into file systems. Files that are actively being updated can be thought of as "live" and present a

---

1 archive *v.* 存档 *n.* 档案文件

2 recover *v.* 重新获得，恢复

3 data retention 数据保持

4 disaster *n.* 灾难，天灾，灾祸

5 reconstitute *v.* 重新组成，重新设立

6 data repository model 数据仓库模型

7 redundancy *n.* 冗余

8 portability *n.* 可携带，轻便

9 compression *n.* 浓缩，压缩

10 deduplication *n.* 数据去重，删除重复数据

11 dry run 演习，排练

12 extraction *n.* 抽出，取出

13 coherent *adj.* 粘在一起的，一致的，连贯的

14 discrete *adj.* 不连续的，离散的

15 file *n.* 文件

challenge to back up. It is also useful to save metadata[1] that describes the computer or the file system being backed up.

Deciding what to back up at any given time is a harder process than it seems. By backing up too much redundant[2] data, the data repository will fill up too quickly. Backing up an insufficient[3] amount of data can eventually lead to the loss of critical information.

(1) Files

1) Copying files. With file-level approach, making copies of files is the simplest and most common way to perform a backup. A means to perform this basic function is included in all backup software and all operating systems.

2) Partial file copying. Instead of copying whole files, one can limit the backup to only the blocks or bytes within a file that have changed in a given period of time. This technique can use substantially less storage space on the backup medium[4], but requires a high level of sophistication to reconstruct[5] files in a restore situation. Some implementations require integration with the source file system.

3) Deleted files. To prevent the unintentional[6] restoration of files that have been intentionally deleted, a record[7] of the deletion must be kept.

(2) File systems

1) File system dump. Instead of copying files within a file system, a copy of the whole file system itself in block-level can be made. This is also known as a raw partition backup and is related to disk imaging[8]. The process usually involves unmounting the file system and running a program like dd (Unix). Because the disk is read sequentially and with large buffers[9], this type of backup can be much faster than reading every file normally, especially when the file system contains many small files, is highly fragmented[10], or is nearly full. But because this method also reads the free disk blocks that contain no useful data, this method can also be slower than conventional reading, especially when the file system is nearly empty. Some file systems provide a "dump" utility that reads the disk sequentially for high performance while skipping unused sections. The corresponding restore utility can selectively restore individual files or the entire volume at the operator's choice.

2) Identification of changes. Some file systems have an archive bit for each file that says it was recently changed. Some backup software looks at the date of the file and compares it with the last backup to determine whether the file was changed.

---

1 metadata *n.* 元数据

2 redundant *adj.* 多余的

3 insufficient *adj.* 不足的，不够的

4 medium *n.* 介质，媒介

5 reconstruct *v.* 重建

6 unintentional *adj.* 不是故意的，无心的

7 record *n.* 记录 *vt.* 记录

8 disk imaging 磁盘镜像

9 buffer *n.* 缓冲区，缓冲器

10 fragmented *adj.* 成碎片的，片断的

3) Versioning file system. A versioning file system keeps track of all changes to a file and makes those changes accessible to the user. Generally this gives access to any previous version, all the way back to the file's creation time. An example of this is the Wayback versioning file system for Linux.

(3) Live data

If a computer system is in use while it is being backed up, the possibility of files being open for reading or writing is real. If a file is open, the contents on disk may not correctly represent what the owner of the file intends. This is especially true for database files of all kinds. The term fuzzy[1] backup can be used to describe a backup of live data that looks like it ran correctly, but does not represent the state of the data at any single point in time. This is because the data being backed up changed in the period of time between when the backup started and when it finished. For databases in particular, fuzzy backups are worthless[2].

1) Snapshot[3] backup. A snapshot is an instantaneous[4] function of some storage systems that presents a copy of the file system as if it were frozen at a specific point in time, often by a copy-on-write[5] mechanism. An effective way to back up live data is to temporarily quiesce them (e.g. close all files), take a snapshot, and then resume live operations. At this point the snapshot can be backed up through normal methods. While a snapshot is very handy for viewing a filesystem as it was at a different point in time, it is hardly an effective backup mechanism by itself.

2) Open file backup. Many backup software packages[6] feature the ability to handle open files in backup operations. Some simply check for openness and try again later. File locking[7] is useful for regulating access to open files.

When attempting to understand the logistics of backing up open files, one must consider that the backup process could take several minutes to back up a large file such as a database. In order to back up a file that is in use, it is vital that the entire backup represent a single-moment snapshot of the file, rather than a simple copy of a read-through.

3) Cold database backup. During a cold backup, the database is closed or locked and not available to users. The data files do not change during the backup process so the database is in a consistent state when it is returned to normal operation.

4) Hot database backup. Some database management systems offer a means to generate a backup image of the database while it is online and usable ("hot"). This usually includes an inconsistent[8] image of the data files plus a log of changes made while the procedure is running. Upon a restore, the changes in the log files are reapplied to bring the copy of the database up-to-

---

1 fuzzy *adj.* 模糊的

2 worthless *adj.* 无价值的，无益的

3 snapshot *n.* 快照

4 instantaneous *adj.* 瞬间的，即刻的，即时的

5 copy-on-write 写时拷贝

6 software package 软件包，程序包

7 file locking 文件锁定

8 inconsistent *adj.* 不一致的，不协调的，矛盾的

date[1] (the point in time at which the initial hot backup ended).

2. Managing the Backup Process

As long as new data are being created and changes are being made, backups will need to be performed at frequent intervals. Individuals and organizations with anything from one computer to thousands of computer systems all require protection of data. The scales may be very different, but the objectives and limitations are essentially the same. Those who perform backups need to know how successful the backups are, regardless of[2] scale.

(1) Objectives

1) Recovery Point Objective (RPO)[3]. The point in time that the restarted infrastructure will reflect. Essentially, this is the roll-back that will be experienced as a result of the recovery. The most desirable RPO would be the point just prior to the data loss event. Making a more recent recovery point achievable requires increasing the frequency of synchronization between the source data and the backup repository.

2) Recovery Time Objective (RTO)[4]. The amount of time elapsed between disaster and restoration of business functions.

(2) Implementation

1) Scheduling. Using a job scheduler[5] can greatly improve the reliability and consistency[6] of backups by removing part of the human element. Many backup software packages include this functionality.

2) Authentication. Over the course of regular operations, the user accounts and/or system agents that perform the backups need to be authenticated at some level. The power to copy all data off of or onto a system requires unrestricted[7] access. Using an authentication mechanism is a good way to prevent the backup scheme from being used for unauthorized activity.

3) Chain of trust. Removable storage media are physical items and must only be handled by trusted individuals. Establishing a chain of trusted individuals (and vendors) is critical to defining the security of the data.

参 考 译 文

数 字 认 证

说到交易，潜在欺诈永远是不容忽视的风险。一个人可以持伪造或变更的文件来冒充他人。

---

1 up-to-date *adj*. 直到现在的，最近的，当代的

2 regardless of 不管，不顾

3 Recovery Point Objective (RPO) 恢复点目标

4 Recovery Time Objective (RTO) 复原时间目标

5 job scheduler 作业安排

6 consistency *n*. 一致性，连贯性

7 unrestricted *adj*. 无限制的，自由的

在线的情况下，一个人也可以不经他人同意，使用其证书，以类似的方式虚构自己的身份。

1. 什么是认证

“认证”是指允许通过电子方式识别自然人或法人的过程。此外，认证还能以电子形式确认数据的来源和完整性，如颁发数字证书以证实网站的真实性。身份验证的总目标是减少欺诈的可能性，特别是当一个人故意虚构自己的身份或未经授权使用另一个人的身份证件时。

“数字认证”或“电子认证”是指以电子方式建立用户身份信任和呈现给信息系统的过程。

由于需要通过网络远程验证个人或实体，这对数字认证的过程提出了技术挑战。其安全级别取决于应用的认证类型、所使用的认证因素以及应用的认证过程。

2. 认证因素

在对在线用户进行身份验证时，可以使用三类因素来确保用户（他或她）就是他们本人。这些因素有以下几类：

- 知识因素——这包括用户的密码、短语、身份证号码（PIN）或要求用户回答预先选择的安全问题。
- 所有权因素——这包括用户拥有的东西，如银行卡、硬件或软件、一次性密码（OTP）令牌或手机。
- 固有特征因素——这些因素与用户的特征或者做的事情有关，并且包括诸如面部、指纹或视网膜图案识别以及其他个人特征的生物特征识别。

3. 认证类型

下面列出了最常用的在线用户认证类型，按照级别的增序来排列：

1）单因素认证——只用三个要素类别中的一个来认证某人的身份。经验表明，单一因素不能提供足够的保护，难以防止恶意入侵和滥用。因此，一旦涉及财务或个人相关交易，应该使用安全性更高的认证。

2）双因素认证（通常称为 2FA），通过把两类不同因素的独立要素组合起来确认用户身份。例如，如果用户使用用户名和密码登录其在线银行账户，并希望完成在线交易，那么除了登录时所用的知识因素（用户名和密码）之外，他还需要输入一个身份验证因素。这个附加因素也必须来自用户名和密码不同的因素类别。网上银行用户通常会使用所有权类别（如 OTP 设备或移动电话）的身份验证机制在文本消息中接收 OTP。OTP 是动态密码，只能使用一次，从而对一系列攻击提供强大的防护等级。

3）多因素认证与 2FA 类似，但可以组合两种以上认证因素来增强安全性，而 2FA 只能使用两种不同的因素。

4）强认证——这种类型认证通常被用作多因素认证或 2FA 的同义词。然而，与多因素认证和 2FA 不同，强认证强制要求不可复制的因素或使用数字证书为用户提供更高级别的认证。如果满足这些标准，多因素身份验证和 2FA 能够提供强大的身份验证。

欧洲中央银行（ECB）将强认证定义为基于三个认证因素中的两个或更多的过程，三个认证因素是：①知识因素；②所有权因素；③固有特征因素。根据欧洲中央银行的定义，为强认证因素选择的个人要素必须是相互独立的，而且至少有一个必须是不可重复使用和不可复制（固有特征除外），也不能通过互联网秘密偷窃。这意味着什么？独立相对容易理解——如果一个组件与另一个组件相关，则攻破一个意味着所有组件都坏了。不可复制性是指时间或用法。使用一次，不能再次使用。例如一次性密码，仅在短时间（如 30 秒）内有效。

在美国，国家安全系统委员会制作的国家信息保证（IA）词汇表类似地定义了强认证，要求多重因素验证和采用先进技术，如使用动态密码或数字证书来验证实体的身份。

4. 注册和认证流程

美国国家标准与技术研究所（NIST）概述了一个相当通用的数字认证模型，可以作为认证过程的基本解释模型，与地理区域或管辖区域无关。

在 NIST 模型中，个人（申请人）向证书服务提供商（CSP）提出申请，从而启动注册过程。一旦 CSP 成功证明了申请人的身份，他或她将成为“订户”，并且在 CSP 和申请人（现在拥有订户角色）之间建立了认证者（如令牌）以及与相应凭证（如用户名）的关联。

CSP 的任务是在证书的整个生命周期内对其进行维护，包括其状态和所有注册数据。订户需要维护认证者。NIST 参考模型的第一部分适用于任何注册过程。在该模型中，还需要进行后续身份验证，例如创建银行账户或在电子政务流程中注册时。

一旦申请人成为“订户”，他或她可以在经过认证的会话中与信任方进行在线交易。在这种交易中，该人担任申请者的角色，向验证者证明拥有一个或多个认证者。验证者和信任方可能是相同的或者两个独立的实体。如果验证者和信任方是分开的，验证者必须向信任方提供关于用户的声明。在这种声明之后，信任方可以启动交易过程。

下面是典型的支付顺序过程：

账户所有者（订户）希望启动交易。他或她首先需要通过一个或多个身份验证者证明，声称账户所有者（申请者他或她）实际上就是其所声称的人（订阅者）。验证由“验证者”完成，验证者在“凭证服务提供商”处验证验证者，验证后向银行（依赖方）的交易部门提供认证声明。在许多银行中，实体“验证者”和“凭证服务提供商”可能也是银行内的实体。

电子政务流程以类似的方式进行（如申请护照）。但在许多情况下，“凭证服务提供商”是外部实体。

5. 结论

欺诈交易给个人、企业和金融机构带来负担，由此造成的损失还有身份盗用的代价都会转嫁给消费者。认证是保护在线交易的一种手段，保护交易双方不被欺诈。

其他概念，如数字签名，也可以利用（强）认证来提高数字交易的安全性、数据完整性和不可否认性。

Unit 10

Text A

The Key Risks Associated with IoT

It's been said that nothing worthwhile is achieved without effort and a certain amount of risk. The Internet of Things(IoT)[1] is most definitely worthwhile and is already the focus of quite a bit of effort, but what about the risks?

All data is at risk these days, not just from hackers and natural disasters, but mechanical failure, human error and sometimes from normal enterprise processes. By extending the data footprint to billions of devices around the planet, however, the number of threat vectors increases dramatically, to the point that conventional security measures such as firewalls are too expensive and too unwieldy to provide adequate protection.

What is the enterprise to do? The first step is to identify the new ways in which the IoT exposes critical assets to risk, and then devise innovative solutions to at least narrow the risk, if not eliminate it altogether. But be forewarned: Not all risks are technological in nature, neither will all of the solutions be.

Here, then, are some of the leading causes of risk, and the means to counter them.

1. Security

The IoT brings with it a wide variety of blind spots that conventional security measures cannot address, says Tim Erlin, senior director of IT security and risk strategy at software developer TripWire. Devices can be assessed for proper security configurations before enterprise resources accept any data, but this is easier said than done. According to a recent company survey, only 30 percent of respondents said they were prepared for security risks in the IoT, while only 34 percent say they can accurately track the number of devices on their networks, let alone the security tools they employ.

Meanwhile, the number of connected devices represents a potentially major escalation in the frequency and intensity of Distributed Denial-of-Service (DDoS)[2] and other types of attacks that harness the power of multiple IP addresses[3] to flood host systems. While emerging IoT infrastructure should provide the dynamic scale needed to accommodate huge increases in traffic, this has yet to be tested in production environments—and the number of connected devices today is only a fraction of what it will be in a few short years.

2. Complexity

The sheer complexity of the IoT is said to be both a blessing and a curse. On the one hand, it is a technological marvel that represents the new height of human ingenuity, on the other hand, it relies on a host of advanced technologies that might not always work exactly as they are supposed to.

One facet of the IoT that is still largely untried is the concept of edge[4] or fog computing[5], in which small, mostly unmanned data centers are networked across regions to provide faster turnaround for data requests. To function properly, these edge systems will have to communicate with numerous devices in its coverage area as well as with other edge systems and with centralized processing centers known as data lakes. Naturally, this requires some fairly sophisticated networking, plus a great deal of coordination between the analytics taking place out on the edge and those in the central data lake[6], which by itself will contain some of the most advanced analytics technologies ever devised.

With all of this cutting-edge technology working in real time, it will probably be quite a while before we see an error-free IoT.

3. Legal

As mentioned above, the IoT creates more than just technological risk; it creates legal risk. According to solicitor Sarah Hall, of U.K. firm Wright Hassall LLP, the IoT affects a number of legal underpinnings surrounding data protection, data sovereignty[7], product liability and a host of other areas. This makes it difficult to determine what laws will apply in a given dispute. Should a driverless car get into an accident, for example, who is liable? The passenger? The owner of the vehicle? The manufacturer? The person who coded the software? Without a clear understanding of how the law will be applied to the IoT, which will only come about through lengthy court processes, the enterprise is open to increasing levels of legal and financial risk as the scale of operations expands.

4. It's Not All Bad

All of this may give the impression that only a madman would embark on an IoT strategy, but the fact is that the same technology that introduces risk can also be used to lessen it.

It's a given that IoT workflows will be so numerous and move so quickly that human operators cannot hope to keep pace with them. That means automation and orchestration[8] will have to play a prominent role in IoT deployments, and increasingly, those solutions are turning to artificial intelligence and cognitive computing to bolster security, availability, data recovery and other functions. As Radware's Carl Herberger noted to TechRadar recently, today's machine learning[9] platforms not only react and respond to threats instantaneously, even proactively, they also adapt themselves to changing attack vectors as they gather more information on normal and abnormal data operations. This will be crucial as the enterprise faces increasingly automated, bot-driven, malware in the IoT.

There is also a growing swell of increasingly sophisticated device management, encryption, access control and other solutions that should make distributed architectures as safe as practicable without inhibiting data and service functionality. A prime example is blockchain, the automated ledger solution originally implemented in digital currency bitcoin, but is now finding its way into a host of applications in which data integrity is paramount.

There is no such thing as a no-risk venture, so the enterprise will have to weigh carefully the risk vs. reward that accompanies every step in the development of IoT infrastructure. And chances are that if any service or application presents too much risk for one organization, it isn't likely to be implemented by anyone else until its concerns are addressed.

In the end, the IoT will only be as risky as the enterprise industry as a whole allows it to be.

New Words

definitely	*adv.* 明确地，干脆地
normal	*n.* 正规，常态
	adj. 正常的，正规的，标准的
vector	*n.* 向量，矢量，带菌者
dramatically	*adv.* 戏剧性地，引人注目地
conventional	*adj.* 惯例的，常规的，传统的
unwieldy	*adj.* 笨拙的，不实用的，难使用的
forewarn	*v.* 预先警告
counter	*n.* 对立面
	v. 反击，还击
	adj. 相反的
survey	*v.* 调查
	n. 测量，调查
escalation	*n.* 扩大，增加
frequency	*n.* 频率，周率，发生次数
intensity	*n.* 强烈，剧烈，强度，亮度
harness	*v.* 利用，控制
accommodate	*v.* 供应，向……提供；使适应，调节；容纳
sheer	*adj.* 全然的，纯粹的，绝对的，彻底的
	v. 避开，躲避，偏航；使避开，使偏航
	adv. 完全，全然
	n. 偏航
marvel	*n.* 奇迹
	v. 大为惊异，觉得惊奇
ingenuity	*n.* 机灵，独创性，精巧，灵活性
untried	*adj.* 未试过的，未经实验的，未经检查的
unmanned	*adj.* 无人的；无人居住的
turnaround	*n.* 转变，转向
coordination	*n.* 协调，调和
devise	*v.* 设计，发明，做出(计划)，想出(办法)

error-free	*n.* 无差错，无错误
risk	*v.* 冒……的危险
	n. 冒险，风险
solicitor	*n.* 律师，法律顾问
underpinning	*n.* 基础，支柱，支撑
sovereignty	*n.* 主权，主权国家
liability	*n.* 责任，义务，债务，负债
dispute	*v. & n.* 争论，辩论，争吵
driverless	*adj.* 无人驾驶的
accident	*n.* 意外事件，事故
liable	*adj.* 有责任的，有义务的
passenger	*n.* 乘客，旅客
vehicle	*n.* 交通工具，车辆
manufacturer	*n.* 制造业者，厂商
madman	*n.* 疯子，精神病患者
numerous	*adj.* 众多的，许多的，无数的
orchestration	*n.* 编排
prominent	*adj.* 卓越的，显著的，突出的
deployment	*n.* 展开，部署
cognitive	*adj.* 认知的，认识的，有感知的
bolster	*v.* 支持
	n. 垫子
instantaneous	*adj.* 瞬间的，即刻的，即时的
swell	*v.* (使)膨胀，增大
practicable	*adj.* 能实行的，行得通的
inhibiting	*adj.* 抑制作用的，约束的
ledger	*n.* 分类账，分户总账，底账
paramount	*adj.* 极为重要的

Phrases

quite a bit	相当多
natural disaster	自然灾害
mechanical failure	机械故障
data footprint	数据占位
blind spot	盲点
a fraction of	一小部分
coverage area	有效区，作用范围
a great deal	大量
data lake	数据湖

cutting-edge technology	前沿技术，尖端技术
a host of	许多，一大群
driverless car	无人驾驶的汽车
embark on	从事，着手
keep pace with	并驾齐驱
Artificial Intelligence(AI)	人工智能
data recovery	数据恢复

Abbreviations

IoT (Internet of Things)	物联网
DDoS (Distributed Denial-of-Service)	分布式拒绝服务

Notes

[1] The Internet of Things (IoT) is the inter-networking of physical devices, vehicles (also referred to as “connected devices” and “smart devices”), buildings, and other items embedded with electronics, software, sensors, actuators, and network connectivity which enable these objects to collect and exchange data.

[2] A Distributed Denial-of-Service (DDoS) attack is an attack in which multiple compromised computer systems attack a target, such as a server, website or other network resource, and cause a denial of service for users of the targeted resource. The flood of incoming messages, connection requests or malformed packets to the target system forces it to slow down or even crash and shut down, thereby denying service to legitimate users or systems.

[3] An Internet Protocol address (IP address) is a numerical label assigned to each device connected to a computer network that uses the Internet Protocol for communication. An IP address serves two principal functions: host or network interface identification and location addressing.

Version 4 of the Internet Protocol (IPv4) defines an IP address as a 32-bit number. However, because of the growth of the Internet and the depletion of available IPv4 addresses, a new version of IP (IPv6), using 128 bits for the IP address, was developed in 1995, and standardized as RFC 2460 in 1998. IPv6 deployment has been ongoing since the mid-2000s.

[4] Edge computing is a method of optimising cloud computing systems by performing data processing at the edge of the network, near the source of the data. This reduces the communications bandwidth needed between sensors and the central datacentre by performing analytics and knowledge generation at or near the source of the data. This approach requires leveraging resources that may not be continuously connected to a network such as laptops, smartphones, tablets and sensors. Edge Computing covers a wide range of technologies including wireless sensor networks, mobile data acquisition, mobile signature analysis, cooperative distributed peer-to-peer ad hoc networking and processing also classifiable as local cloud/fog computing and grid/mesh computing, dew computing, mobile edge computing, cloudlet, distributed data storage and retrieval, autonomic

self-healing networks, remote cloud services, augmented reality, and more.

[5] Fog computing or fog networking, also known as fogging, is an architecture that uses one or more collaborative end-user clients or near-user edge devices to carry out a substantial amount of storage (rather than stored primarily in cloud data centers), communication (rather than routed over the Internet backbone), control, configuration, measurement and management (rather than controlled primarily by network gateways such as those in the LTE core network).

[6] A data lake is a method of storing data within a system or repository, in its natural format, that facilitates the collocation of data in various schemata and structural forms, usually object blobs or files. The idea of data lake is to have a single store of all data in the enterprise ranging from raw data (which implies exact copy of source system data) to transformed data which is used for various tasks including reporting, visualization, analytics and machine learning. The data lake includes structured data from relational databases (rows and columns), semi-structured data (CSV, logs, XML, JSON), unstructured data (e-mails, documents, PDFs) and even binary data (images, audio, video) thus creating a centralized data store accommodating all forms of data.

A data swamp is a deteriorated data lake, that is inaccessible to its intended users and that provides very little value to the company.

[7] Data sovereignty is the concept that information which has been converted and stored in binary digital form is subject to the laws of the country in which it is located.

Many of the current concerns that surround data sovereignty relate to enforcing privacy regulations and preventing data that is stored in a foreign country from being subpoenaed by the host country's government.

[8] Orchestration is the automated arrangement, coordination, and management of computer systems, middleware, and services.

[9] Machine learning is a type of Artificial Intelligence (AI) that allows software applications to become more accurate in predicting outcomes without being explicitly programmed. The basic premise of machine learning is to build algorithms that can receive input data and use statistical analysis to predict an output value within an acceptable range.

Machine learning algorithms are often categorized as being supervised or unsupervised. Supervised algorithms require humans to provide both input and desired output, in addition to furnishing feedback about the accuracy of predictions during training. Once training is complete, the algorithm will apply what was learned to new data. Unsupervised algorithms do not need to be trained with desired outcome data. Instead, they use an iterative approach called deep learning to review data and arrive at conclusions. Unsupervised learning algorithms are used for more complex processing tasks than supervised learning systems.

Exercises

[Ex. 1] Answer the following questions according to the text.

1. Where does the risk all data is at come from these days?
2. What is the enterprise to do with the risk?

3. What does a recent company survey show?

4. Why is the sheer complexity of the IoT said to be both a blessing and a curse?

5. What is one facet of the IoT that is still largely untried? What will these edge systems do to function properly?

6. What does the IoT do according to solicitor Sarah Hall?

7. What does the author want to tell us by the example "Should a driverless car get into an accident, for, who is liable? The passenger? The owner of the vehicle? The manufacturer? The person who coded the software?"

8. What is blockchain?

9. What will the enterprise do since there is no such thing as a no-risk venture?

10. What will happen if any service or application presents too much risk for one organization?

[Ex. 2] Translate the following terms or phrases from English into Chinese and vice versa.

1. data lake	1. ______
2. Artificial Intelligence(AI)	2. ______
3. a fraction of	3. ______
4. coverage area	4. ______
5. data recovery	5. ______
6. 惯例的，常规的， 传统的	6. ______
7. 展开，部署	7. ______
8. 频率，周率，发生次数	8. ______
9. 制造业者，厂商	9. ______
10.向量，矢量	10. ______

[Ex. 3] Translate the following passage into Chinese.

The Internet of Things—This Is Where We Are Going

In one vision of the future, every "thing" is connected to the Internet. This "Internet of Things" will bring about revolutionary change in how we interact with our environment and, more importantly, how we live our lives.

The idea of everything being connected to the Internet is not new, but it's increasingly becoming a reality. The Internet of Things came into being in 2008 when the number of things connected to the Internet was greater than the number of people who were connected.

The technical Utopians have portrayed the Internet of Things as a good thing that will bring untold benefits. They are supported by all the companies that stand to benefit by the increasing connectedness of everything.

Universal connectivity, sensors and computers that are able to collect, analyse and act on this data will bring about improvements in health and food production. In a roundabout way, it might even alleviate poverty.

On the other side are the sceptics who warn of the dangers inherent in not only having an ever growing Internet of Things, but our increasing reliance on it.

[Ex. 4] Fill in the blanks with the words given below.

upgrade	minimum	make	install	unique
discover	synchronize	smart	download	passwords

1. Keep Your Software Updated

The best defense against hackers is keeping all your software patched. It's an arms race with the hackers to ___1___ security breaches before the vendors do. That's why it's important to keep all your software as up to date as possible.

That's why Windows 10 requires users to actually ___2___ updates for the home versions, which hasn't necessarily been a popular decision. Business users can hold off actually installing them, so they can ___3___ sure the updates don't break anything, but they'll have to install the updates eventually. The faster update cadence in Windows 10 is an attempt to keep up, with weekly cumulative updates.

While Linux distributions offer various package managers and Ubuntu offers business users a paid tool to ___4___ updates across fleets of computers, most businesses are going to be running Windows and are only going to wait to ___5___ to Windows 10 until the bugs are completely ironed out.

That makes the use of third-party tools to manage software updates important. One possible tool is Ninite Pro. While Ninite lets ordinary users ___6___, install and update a range of popular applications, Ninite Pro lets businesses manage updates across large numbers of computers. Large organizations like NASA and Tupperware already rely on it.

Windows Server Update Services (WSUS) let you roll out Windows updates from Windows Server installations to Windows desktops.

2. Use a Password Manager

While using strong and ___7___ passwords ranks up there with working smoke detectors as an essential practice, it's hard to remember all of the passwords for the different accounts most people have. Using good passwords is something that both expert and non-expert users have in common, but ___8___ users know when they can automate tedious processes. Password managers are a good example.

There have been several tools to help users keep strong passwords with a ___9___ of effort. The most well known is LastPass. While LastPass is best known as a consumer-based application, the company offers a version for enterprise designed to support large organizations. LastPass attempts to create truly randomized ___10___ while users only have to remember one login. Even though LastPass had a well-publicized breach earlier this year, it uses strong encryption. The passwords are only decrypted on the local machine, so having access to the password vault is virtually useless for an attacker.

Text B

Beware! Your Devices Are Spying on You

They're here, they're there and they're everywhere—recording your every move. Listening to your every word and listing your every interaction. You invited them in, gave them access to your space through cell phones, wearables[1], webcams and basically everything that connects to the Internet. It is a pretty fantastical thought and a bit scary. Yet it hasn't seemed to sway the millions of people away from interacting with these listening devices, gobbling them up by the hundreds of millions. It is hard not to get pulled into the hype... and the convenience. These gateway devices are our portals to the Internet of Things (IoT) and are paving the way to make our lives "easier." As we continue to open ourselves and our families to this monitoring, it is imperative that we recognize our exposures. Here are a few devices and thoughts to consider.

1. Baby Monitors

A baby monitor is supposed to make us feel safe. Recently they have come out with video baby monitors. Now we are able to watch our little ones from anywhere. Not only is this cute, but you can distinguish if a whimper is something that needs attending to or if your darling is merely dreaming. In recent years, these monitors have taken even bigger steps: You can scan the room from anywhere. At work, do you want to see if the sitter put your baby to bed? No problem—open your app and take a peek. Sounds like it should bring you a big sense of relief... or does it? How strong is that password securing this video device? A hacker can also watch your baby at home and learn the routines of your family. Make sure to secure your routers and modems.

2. Garage Door Openers and Automatic Locks

Remember when the only option we had to secure our garage was with a metal key? I certainly do. Today we are able to lock and unlock our homes with our smartphones and similar devices—from a different state let alone from across town. Now while this is convenient in many situations, the potential for misuse is alarming. If you can open your garage door, so can hackers if your system is not properly safeguarded. Seventy-three percent of adults are victims of cyber-crimes each year. It isn't too hard to imagine someone breaking into an improperly secured home.

3. Kinect[2] for Xbox

Xbox is amazing. It allows users to talk to each other and video chat. Users can allow Kinect to identify their face and enable automatic sign-in to an Xbox Live profile. If you are playing with a total stranger, you are letting them into your space. That is obvious. But quite literally the spies are watching you as well. Microsoft's Kinect has been linked with Britain's spy agency known as GCHQ, and the snooping doesn't end there. Kinect has been mining data for our own National Security Agency for over a decade. Think of the big data trove being created by today's teenagers.

The NSA now has your child's image, most probably yours, and that of your living room. Openly transmitted information is gleaned from your camera, your chat logs, audio, and it is all linked to your IP address. Talk about the Internet of Things! Remember to watch your back. Secure your network and shut down your game consoles when not in use.

4. Smart TV

The television has changed a great deal since its inception—from the dial, to the remote, to today, where we can use voice command features. While I am still of the age where I can manage to use a traditional remote, I do appreciate the potential that voice-activated technology provides in today's smart TVs. However I have chosen not to let my television eavesdrop and record my family's private conversations. Keep in mind that when a voice request is processed, smart TVs are programmed to send this data to third parties. I don't want my personal information sent to just anyone. While the TVs don't monitor before they hear key words, they are listening for triggers and once they hear them, they are "aware." And, if it isn't bad enough to keep your privacy from legit companies, what would happen if a hacker broke in and could listen to your private moments? Do your homework and know your TV's security and privacy options.

5. Webcams

Early on, a webcam was an upgrade in a computer or device. Today webcams are something that every business person is expected to have. Webcams have many benefits. Prior to webcams and video conferencing we traveled around the world for brief meetings. Now, with a double click or a button pushed we can meet face-to-face with individuals anywhere. It is amazing. However, these same devices have also given hackers worldwide the ability to enter our affairs and look around without ever being noticed. Sometimes it is voyeuristic, or sometimes it is malicious. Imagine not shutting down your webcam, leaving your computer on, and getting undressed. A hacker has taken over your webcam. You don't know it, but you are being watched. Shut down and secure your devices after each use.

6. Protect Yourself

After reading all the above, it may seem I am paranoid. As a risk manager I question most everything and then use a reverse process to dissect, plan and mitigate. While the manufacturers of listening devices have a moral responsibility to their users, you can take charge and intercede on your own behalf. There are specific ways to protect yourself with each device. First and foremost, use complex passwords. Use a password generator. Companies including Dashlane.com and Keepass.info offer free password generators and online password vaults that are very effective in protecting you and your personal information from the point of login. Know your vendor, research your device and understand the security and privacy settings available. Stay tuned to news about the technology you own and take the time to reach out to an area IT or home security expert to help close any open windows or portals. Don't just pause the feature. Always turn off the device once you are finished using it. If you are at home, don't use the online feature of the baby monitor. This way your routine won't be as obvious.

Be smart and be careful. Remember, the online world has become a dangerous place, yet with proper planning you can know your risks and work to reduce them.

New Words

wearable	*adj*. 可穿戴的，可佩带的
webcam	*n*. 网络摄像机，网络摄像头
fantastical	*adj*. 奇异的，幻想的，异想天开的；极大的；极好的，精彩的
scary	*adj*. 引起惊慌的
sway	*v*. 摇摆，摇动
hype	*n*. 大肆宣传，大做广告
convenience	*n*. 便利，方便
gateway	*n*. 门，通路，网关
portal	*n*. 入口，门户
imperative	*n*. 命令；需要；规则
	adj. 命令的；强制的
video	*n*. 录像，视频
whimper	*v*. 呜咽，哀诉
	n. 啜泣声，呜咽声
scan	*v*. 扫描，细看，审视，浏览
	n. 扫描
sitter	*n*. 代人临时看管小孩的人
garage	*n*. 车库，汽车间，修车厂
	v. 放入车库
alarming	*adj*. 令人担忧的
amazing	*adj*. 令人惊异的
snoop	*v*. 探听，调查，偷窃
	n. 到处窥视、爱管闲事的人；私家侦探
trove	*n*. 被发现的东西，收藏的东西
glean	*v*. 收集
audio	*adj*. 音频的，声频的，声音的
inception	*n*. 起初
legit	*adj*. 合法的
conference	*n*. 会议，讨论会，协商会
voyeuristic	*adj*. 喜好窥探他人隐私的
undressed	*adj*. 穿着便服的，裸体的
paranoid	*n*. 妄想症患者
	adj. 妄想狂的，多疑的
dissect	*v*. 切开仔细研究，把……切成碎片
moral	*adj*. 道德(上)的，精神的
	n. 寓意，教训

intercede v. 调解

Phrases

gobble sb./sth. up 狼吞虎咽地吃掉，贪婪地吃掉
get pulled into 被拉进
gateway device 网关设备
be supposed to 应该，被期望
come out with 发表，公布，说出，展出，供应
take a peek 匆匆一窥
video chat 视频聊天
Smart TV 智能电视
voice-activated technology 语音开机技术
take charge 看管，负责
password generator 密码生成器

Abbreviations

GCHQ (Government Communications Headquarters) 国家通信总局

Notes

[1] The terms "wearable technology", "wearable devices", and "wearables" all refer to electronic technologies or computers that are incorporated into items of clothing and accessories which can comfortably be worn on the body. These wearable devices can perform many of the same computing tasks as mobile phones and laptop computers; however, in some cases, wearable technology can outperform these hand-held devices entirely. Wearable technology tends to be more sophisticated than hand-held technology on the market today because it can provide sensory and scanning features not typically seen in mobile and laptop devices, such as biofeedback and tracking of physiological function.

[2] Kinect (codenamed Project Natal during development) is a line of motion sensing input devices by Microsoft for Xbox 360 and Xbox One video game consoles and Microsoft Windows PCs. Based around a webcam-style add-on peripheral, it enables users to control and interact with their console/computer without the need for a game controller, through a natural user interface using gestures and spoken commands.

Exercises

[Ex. 5] Fill in the following blanks according to the text.

1. A baby monitor is supposed to ________________. Recently they have come out with ________________. Now we are able to ________________________________.

2. A hacker can also watch your baby ____________ and learn ___________________. Make sure to secure_______________________.

3. Today we are able to lock and unlock our homes ________________ and similar devices —from ________________, let alone from across town.

4. If you can open your garage door, so can hackers if your system ________________. ________________ are victims of cyber-crimes each year.

5. Xbox is amazing. It allows users to ________________ and video chat. Users can allow Kinect to ________________ and ________________ to an Xbox Live profile.

6. The television has changed a great deal since its inception—from ________________, to ________________, to today, where we can ________________.

7. Early on, a webcam was ________________ in a computer or device. Today webcams are something that every business person ________________.

8. Prior to webcams and video conferencing we________________ for brief meetings. Now, with ________________ or a button pushed we can ________________ with individuals anywhere.

9. There are ________________ to protect yourself with each device. First and foremost, ________________. Use a password generator.

10. In the last paragraph, the author tells us to ________________ and ________________. We should remember that the online world has become ________________, yet with ________________ we can know our risks and ________________.

Reading Material

Internet of Things: Who Owns the Data?

Smart cars, connected health, smart grids[1], smart cities—the world is becoming connected in a way that was the territory[2] of science fiction[3] just a few short years ago. Ericsson and Cisco both predict upwards of 50 billion devices will be connected to the Internet by 2020 in a network of "things" that will extend well beyond smartphones, laptops and game consoles[4] to scanners, sensors[5], etc. This promise of the Internet of Things will increase today's data load factors by several orders of magnitude[6]. While this creates questions about how data is collected, ingested[7], stored and queried, one of the most important considerations will be around ownership and governance around that data.

1. Constant Data Collection

There is already friction[8] between consumers and online services around ownership of data

1 smart grid 智能电网
2 territory *n.* 领土，版图，地域
3 science fiction 科幻小说
4 console *n.* 控制台
5 sensor *n.* 传感器
6 order of magnitude 数量级
7 ingest *v.* 摄取，吸收
8 friction *n.* 摩擦，摩擦力

collected from the likes of Facebook, Google, Twitter and others. There have been flare-ups of backlash against Facebook, for example, which claims ownership of your photos and content posted to your personal news feeds. Most consumers don't realize that when they sign Facebook's terms of agreement without reading them. Consumers, however, are starting to understand the implications of giving up personal information online as they start to see targeted advertising based on their online profiles and behaviors.

For the most part, as we have seen from this increase in behavior-based, targeted advertising, the access and use of this data is primarily driven by money: Ad networks generate revenue from advertisers through more targeted programs that in turn generate revenue for advertisers by getting consumers to spend more money because of more targeted advertising—the circle of life, if you will. Maybe you like the enhanced[1] experience Yahoo! gives you for fantasy[2] football, but what happens when there are 50 billion connected devices, the majority of which are machines like sensors embedded in cars, clothes, cardiac monitors[3] and more?

2. Who's in Control?

As data is increasingly collected and shared, the most important question—at least for consumers—is who owns the data in your smart meter[4] and what does that information tell you, or tell others about you? If combining data from smart cars with data from smart traffic grids[5] and smart energy delivery has value, how do these systems know how to speak to one another, and what governs who can access this data and how? What about medical data? When the sensor sewn into a piece of clothing, or on a wristband that is tracking vital signs and alerting[6] your doctor when certain thresholds are breached, where and how is that data held and managed?

Regulators have not been idle on the data ownership and protection front. MIT professor Alex (Sandy) Pentland has written extensively about privacy, data ownership and data control.

"You can imagine using big data to make a world that is incredibly invasive[7], incredibly Big Brother...George Orwell was not nearly creative enough when he wrote '1984'."

Pentland has led several sessions at the World Economic Forum, which culminated in the chairman of the Federal Trade Commission putting forward the U.S. Consumer Privacy Bill of Rights, and the EU introducing stringent[8] (and controversial[9]) laws forcing businesses to embed data protection.

Unfortunately, data ownership and privacy legislation[10] is by no means a done deal. In early

---

1 enhanced *adj.* 增强的，提高的，放大的
2 fantasy *n.* 幻想，白日梦
3 cardiac monitor 心电监护仪
4 meter *n.* 仪表
5 smart traffic grid 智能交通网
6 alert *v.* 发信号
7 invasive *adj.* 入侵的
8 stringent *adj.* 严厉的，迫切的
9 controversial *adj.* 争论的，争议的
10 legislation *n.* 立法，法律的制定(或通过)

June, the National Telecommunications and Information Administration issued a Request for Comments on how issues raised by big data impact the Consumer Privacy Bill of Rights.

The primary focus though, is to put individuals in control of their own data and how it is used, and to ensure that their data is safeguarded[1]. While some businesses may look at this as limiting the success of the Internet of the Things (from a revenue perspective), it is absolutely[2] critical to the realization[3] of a wide-scale adoption.

3. Data Issues To Be Addressed

While ownership of data may still be up for grabs[4], companies will also need to grapple with[5] many other considerations:

- Who is the steward[6] of the data? Facebook may well be the owner, but it will try to allow the user to, within limits, provide the stewardship.
- How is the data accessed? Will it be pushed into an all-access portal, or only through a secure API[7]?
- How is the data defined, literally? The socialization[8] and exposure of data can be significantly impacted by the exact definitions used and the drivers behind those definitions (multiple standards-based approaches).

What security measures are in place and who administers[9] them? The security administrator may allow access in unpopular[10] ways. However, they may also have great policies in place that are ineffective[11] based on weak execution. We all recall Heartbleed.

- Who owns the derivative[12] information about the data? This is a more nuanced consideration around ownership of emergent patterns identified in data, and the implications of those patterns.

Ultimately, enterprises, advertisers[13] and others will need to prove that the value they will deliver to the consumer will be worth that consumer giving them information about themselves and, that the consumer can trust their information to be safe. In the end, the consumers should make that decision. While regulators have a significant part to play in forcing the issue of data protection, there is responsibility on all fronts. A combination of industry voices, large players in the market, and dare I say it, government, will be well served to work together to strike the right balance.

---

1 safeguard *v.* 维护，保护，捍卫
2 absolutely *adv.* 完全地，绝对地
3 realization *n.* 实现
4 be up for grabs 供人获取，赢得，竞争
5 grapple with 努力解决；抓住
6 steward *n.* 管理者，管家
7 API 应用编程接口(Application Programming Interface)
8 socialization *n.* 社会化
9 administer *v.* 管理，执行
10 unpopular *adj.* 不受欢迎的
11 ineffective *adj.* 无效的，(指人)工作效率低的
12 derivative *adj.* 引出的，派生的
13 advertiser *n.* 登广告者，广告客户

Time will tell.

参 考 译 文

与物联网相关的主要风险

据说，如果不努力，不冒一定的风险就不会获得有价值的东西。物联网（IoT）绝对有价值，也已经成为人们努力的焦点，但是风险呢？

现在所有数据都处于风险之中，而这些风险不仅仅来自黑客和自然灾害，也来自机械故障、人为错误，有时还会来自正常的企业流程。然而，地球上数十亿设备上都有数据，这使威胁向量的数量急剧增加，以至于传统的安全措施（如防火墙）变得过于昂贵也难以提供足够的保护。

企业要做什么？第一步是确定物联网将关键资产暴露于风险的新方式，然后设计创新解决方案，如果不能完全消除风险起码要缩小风险。但是要预先警告：不是所有的风险都是技术性的，也不是所有的解决方案都是技术性的。

那么，此处列出产生风险的一些主要原因以及抗风险的手段：

1. 安全性

软件开发商 TripWire 的 IT 安全和风险策略高级总监蒂姆·厄林表示，IoT 带来了传统安全措施无法解决的各种盲点。在企业资源接受任何数据之前，可以对设备进行适当的安全配置评估，但这说起来容易做起来难。根据最近的公司调查，只有 30%的受访者表示他们为物联网的安全风险做好了准备，只有 34%的受访者表示可以准确地跟踪其网络上的设备数量，更不用说他们所使用的安全工具了。

同时，连接设备的数量表明了潜在分布式拒绝服务（DDoS）和其他类型攻击的频率和强度，这些攻击能利用多个 IP 地址来淹没主机系统。虽然新兴的 IoT 基础设施应该提供所需的动态规模，以适应业务量的巨大增长，但在生产环境中尚未得到测试，而今天连接的设备数量远远小于未来几年的设备数量。

2. 复杂性

物联网的复杂性被认为既是福也是祸。一方面，它代表了人类创造力的新高度，是一种技术奇迹；但另一方面，它依赖于许多先进的技术，而这些技术可能并不总是确切地按照应有的方式工作。

IoT 的一个方面仍然是未经考验的，即边缘或“雾”计算的概念，其中小型的大多数无人数据中心跨区域联网，为数据请求提供更快的传送时间。为了正常运行，这些边缘系统必须与其覆盖区域中的许多设备以及其他边缘系统和称为数据湖的集中处理中心进行通信。当然，这需要在边缘分析和中央数据湖中的分析之间进行大量协调，而协调本身也包括一些最先进的分析技术。

随着所有这些尖端技术的实时工作，可能会需要一段时间才能看到无差错的物联网。

3. 法律

如上所述，物联网不仅造成了技术风险，也造成了法律风险。据英国威廉·哈索尔律师事务所的律师莎拉·霍尔介绍，物联网影响了包括数据保护、数据主权、产品责任以及许多其他领域的若干法律基础。这使得在特定争议中很难确定适用哪些法律。例如，如果一辆无人驾驶汽车发生事故，谁负责？乘客？车主？生产厂家？编软件的人？如果不能清楚地了解如何将法律应用于物联网，就只能通过漫长的法庭程序来实现。随着业务规模的扩大，企业可能面临越来越多的法律和财务风险。

4. 并非有害无益

这一切可能会给人这样的印象：只有疯子才会着手实施物联网战略，但事实是，带来风险的技术也可能用来减少风险。

物联网工作流如此众多、发展得如此之快，以至于人类操作者跟不上其步伐。这意味着自动化和业务流程编排必将在物联网部署中发挥突出作用，越来越多的解决方案正在转向人工智能和认知计算，以加强安全性、可用性、数据恢复和其他功能。Radware 的卡尔·赫伯格最近向 TechRadar 表示，今天的机器学习平台不仅可以立即对威胁做出反应和响应，甚至可以主动响应，因为其收集了更多的正常和异常数据操作信息，它们也会适应不断变化的攻击向量。因为企业在物联网中面临越来越多的自动的、机器人驱动的恶意软件，所以这就至关重要。

越来越复杂的设备管理、加密、访问控制和其他解决方案也在不断增长，这些解决方案应该使分布式架构尽可能安全，而不会妨碍数据和服务功能。一个主要的例子是区块链，它是自动分类解决方案，最初以数字货币比特币实现，但现在正在进入数据完整性至关重要的一系列应用程序。

没有什么是没有风险的，所以企业必须仔细衡量与物联网基础设施开发相伴的每一步的风险与回报。如果任一服务或应用程序对组织可能造成太大的风险，在解决问题之前，其他人都不能实施它。

最终，物联网的风险取决于整个行业允许它所具有的风险。

Unit 11

Text A

Cloud Computing Security

Cloud computing security or, more simply, cloud security is an evolving sub-domain of computer security, network security, and, more broadly, information security. It refers to a broad set of policies, technologies, and controls deployed to protect data, applications, and the associated infrastructure of cloud computing.

1. Security Issues Associated with the Cloud Computing

Cloud computing and storage solutions provide users and enterprises with various capabilities to store and process their data in third-party data centers. Organizations use the cloud in a variety of different service models (SaaS[1], PaaS[2], and IaaS[3]) and deployment models (Private, Public, Hybrid, and Community). There are a number of security concerns associated with cloud computing. These issues fall into two broad categories: security issues faced by cloud providers (organizations providing software-, platform-, or infrastructure-as-a-service via the cloud) and security issues faced by their customers (companies or organizations who host applications or store data on the cloud). The responsibility is shared, however. The provider must ensure that their infrastructure is secure and that their clients' data and applications are protected, while the user must take measures to fortify their application and use strong passwords and authentication measures.

When an organization elects to store data or host applications on the public cloud, it loses its ability to have physical access to the servers hosting its information. As a result, potentially sensitive data is at risk from insider attacks. According to a recent Cloud Security Alliance[4] Report, insider attacks are the sixth biggest threat in cloud computing. Therefore, Cloud Service providers must ensure that thorough background checks are conducted for employees who have physical access to the servers in the data center. Additionally, data centers must be frequently monitored for suspicious activity.

In order to conserve resources, cut costs, and maintain efficiency, cloud service providers often store more than one customer's data on the same server. As a result, there is a chance that one user's private data can be viewed by other users (possibly even competitors). To handle such sensitive situations, cloud service providers should ensure proper data isolation and logical storage segregation.

The extensive use of virtualization[5] in implementing cloud infrastructure brings unique security concerns for customers or tenants of a public cloud service. Virtualization alters the relationship between the OS and underlying hardware—be it computing, storage or even networking. This introduces an additional layer—virtualization—that itself must be properly configured, managed and secured. Specific concerns include the potential to compromise the virtualization software, or hypervisor[6]. While these concerns are largely theoretical, they do exist. For example, a breach in the administrator workstation with the management software of the virtualization software can cause the whole data center to go down or be reconfigured to an attacker's liking.

2. Cloud Security Controls

Cloud security architecture is effective only if the correct defensive implementations are in place. An efficient cloud security architecture should recognize the issues that will arise with security management. The security management addresses these issues with security controls. These controls are put in place to safeguard any weaknesses in the system and reduce the effect of an attack. While there are many types of controls behind a cloud security architecture, they can usually be found in one of the following categories.

(1) Deterrent Controls

These controls are intended to reduce attacks on a cloud system. Much like a warning sign on a fence or a property, deterrent controls typically reduce the threat level by informing potential attackers that there will be adverse consequences for them if they proceed. (Some consider them a subset of preventive controls.)

(2) Preventive Controls

Preventive controls strengthen the system against incidents, generally by reducing if not actually eliminating vulnerabilities. Strong authentication of cloud users, for instance, makes it less likely that unauthorized users can access cloud systems, and more likely that cloud users are positively identified.

(3) Detective Controls

Detective controls are intended to detect and react appropriately to any incidents that occur. In the event of an attack, a detective control will signal the preventative or corrective controls to address the issue. System and network security monitoring, including intrusion detection and prevention arrangements, are typically employed to detect attacks on cloud systems and the supporting communications infrastructure.

(4) Corrective Controls

Corrective controls reduce the consequences of an incident, normally by limiting the damage. They come into effect during or after an incident. Restoring system backups in order to rebuild a compromised system is an example of a corrective control.

3. Dimensions of Cloud Security

It is generally recommended that information security controls be selected and implemented according and in proportion to the risks, typically by assessing the threats, vulnerabilities and impacts.

4. Security and Privacy

(1) Identity Management

Every enterprise will have its own identity management system to control access to information and computing resources. Cloud providers either integrate the customer's identity management system into their own infrastructure, using federation or SSO[7] technology, or a biometric-based identification system, or provide an identity management solution of their own. CloudID, for instance, provides a privacy-preserving cloud-based and cross-enterprise biometric identification solutions for this problem. It links the confidential information of the users to their biometrics and stores it in an encrypted fashion. Making use of a searchable encryption technique, biometric identification is performed in encrypted domain to make sure that the cloud provider or potential attackers do not gain access to any sensitive data or even the contents of the individual queries.

(2) Physical Security

Cloud service providers physically secure the IT hardware (servers, routers, cables etc.) against unauthorized access, interference, theft, fires, floods etc. and ensure that essential supplies (such as electricity) are sufficiently robust to minimize the possibility of disruption. This is normally achieved by serving cloud applications from world-class (i.e. professionally specified, designed, constructed, managed, monitored and maintained) data centers.

(3) Personnel Security

Various information security concerns relating to the IT and other professionals associated with cloud services are typically handled through pre-, para- and post-employment activities such as security screening potential recruits, security awareness and training programs.

(4) Privacy

Providers ensure that all critical data (credit card numbers, for example) are masked or encrypted and that only authorized users have access to data in its entirety. Moreover, digital identities and credentials must be protected as should any data that the provider collects or produces about customer activity in the cloud.

5. Data Security

There are a number of security threats associated with cloud data services, not only covering traditional security threats, e.g., network eavesdropping, illegal invasion, and denial of service attacks, but also including specific cloud computing threats, e.g., side channel attacks[8], virtualization vulnerabilities, and abuse of cloud services. To throttle the threats the following security requirements are to be met in a cloud data service.

(1) Data Confidentiality

Data confidentiality is the property that data contents are not made available or disclosed to illegal users. Outsourced data is stored in a cloud and out of the owners' direct control. Only authorized users can access the sensitive data while others should not gain any information of the data. Meanwhile, data owners expect to fully utilize cloud data services, e.g., data search, data computation, and data sharing, without the leakage of the data contents to other adversaries.

(2) Data Access Controllability

Access controllability means that a data owner can perform the selective restriction of access to his data outsourced to cloud. Legal users can be authorized by the owner to access the data, while others cannot access it without permissions. Further, it is desirable to enforce fine-grained access control to the outsourced data, i.e., different users should be granted different access privileges with regard to different data pieces. The access authorization must be controlled only by the owner in untrusted cloud environments.

(3) Data Integrity

Data integrity demands maintaining and assuring the accuracy and completeness of data. A data owner always expects that his data in a cloud can be stored correctly and trustworthily. It means that the data should not be illegally tampered, improperly modified, deliberately deleted, or maliciously fabricated. If any undesirable operations corrupt or delete the data, the owner should be able to detect the corruption or loss. Further, when a portion of the outsourced data is corrupted or lost, it can still be retrieved by the data users.

New Words

sub-domain	*n*.	子域
broadly	*adv*.	广泛地
deploy	*v*.	展开；配置
hybrid	*n*.	混合物
	adj.	混合的
community	*n*.	团体；共有，一致；共同体；(生物)群落
fortify	*v*.	设防于，筑防御工事
elect	*v*.	选择
conserve	*v*.	保存
efficiency	*n*.	效率
competitor	*n*.	竞争者
handle	*v*.	处理，解决
ensure	*v*.	确保
segregation	*n*.	隔离，分离
virtualization	*n*.	虚拟化
unique	*adj*.	唯一的，独特的
tenant	*n*.	租户
hypervisor	*n*.	超级管理程序，虚拟机管理程序
workstation	*n*.	工作站
liking	*n*.	爱好，嗜好
deterrent	*n*.	威慑
inform	*v*.	告知
adverse	*adj*.	不利的，有害的；逆的，相反的

consequence	*n.*	结果，后果
preventive	*adj.*	预防性的
strengthen	*v.*	加强，巩固
preventative	*adj.*	预防性的
corrective	*adj.*	纠正的，矫正的
rebuild	*v.*	重建，复原
dimension	*n.*	尺寸，尺度，维(数)，度(数)，元
integrate	*v.*	整合，使成整体，使一体化
federation	*n.*	联合，联盟
interference	*n.*	冲突，干涉
screening	*n.*	筛查
	v.	审查
recruit	*n.*	新人，新分子，新会员
masked	*v.*	伪装
	adj.	戴面具的
invasion	*n.*	入侵
throttle	*v.*	扼杀
leakage	*n.*	漏，泄漏，渗漏
controllability	*n.*	可控性
fabricate	*v.*	捏造，伪造

Phrases

cloud computing	云计算
third-party data center	第三方数据中心
insider attack	内部攻击
cut cost	消减成本，降低成本
data isolation	数据隔离
to one's liking	合……的胃口，合……的意
public cloud	公共云
warning sign	警报信号，警告，警告标志
in proportion to…	与……成比例
identity management system	身份管理系统
biometric-based identification system	基于生物特征的身份识别系统
Side Channel Attack(SCA)	边信道攻击，又称侧信道攻击
outsourced data	外包数据
fine-grained access control	精细化访问控制
with regard to	关于

Abbreviations

SaaS (Software as a Service)	软件即服务
PaaS (Platform as a Service)	平台即服务
IaaS (Infrastructure as a Service)	基础设施即服务
OS (Operating System)	操作系统
SSO (Single Sign-On)	单点登录

Notes

[1] Software as a Service (SaaS) is a software licensing and delivery model in which software is licensed on a subscription basis and is centrally hosted. It is sometimes referred to as "on-demand software", and was formerly referred to as "software plus services" by Microsoft. SaaS is typically accessed by users using a thin client via a web browser. SaaS has become a common delivery model for many business applications, including office and messaging software, payroll processing software, DBMS software, management software, CAD software, development software, gamification, virtualization, accounting, collaboration, Customer Relationship Management (CRM), Management Information Systems (MIS), Enterprise Resource Planning (ERP), invoicing, Human Resource Management (HRM), talent acquisition, Content Management (CM), and service desk management. SaaS has been incorporated into the strategy of nearly all leading enterprise software companies.

[2] Platform as a Service (PaaS) or Application Platform as a Service (aPaaS) is a category of cloud computing services that provides a platform allowing customers to develop, run, and manage applications without the complexity of building and maintaining the infrastructure typically associated with developing and launching an app. PaaS can be delivered in two ways: As a public cloud service from a provider, where the consumer controls software deployment with minimal configuration options, and the provider provides the networks, servers, storage, Operating System (OS), middleware (e.g. Java runtime, .NET runtime, integration, etc.), database and other services to host the consumer's application; or as a private service (software or appliance) inside the firewall, or as software deployed on a public infrastructure as a service.

[3] Infrastructure as a Service (IaaS) is a form of cloud computing that provides virtualized computing resources over the Internet. IaaS is one of three main categories of cloud computing services, alongside Software as a Service (SaaS) and Platform as a Service (PaaS).

[4] Cloud Security Alliance (CSA) is a not-for-profit organization with a mission to "promote the use of best practices for providing security assurance within Cloud Computing, and to provide education on the uses of Cloud Computing to help secure all other forms of computing".

[5] In computing, virtualization refers to the act of creating a virtual (rather than actual) version of something, including virtual computer hardware platforms, storage devices, and computer network resources.

[6] A hypervisor or Virtual Machine Monitor (VMM) is computer software, firmware or

hardware that creates and runs virtual machines. A computer on which a hypervisor runs one or more virtual machines is called a host machine, and each virtual machine is called a guest machine. The hypervisor presents the guest operating systems with a virtual operating platform and manages the execution of the guest operating systems.

[7] SSO (single sign-on) is a property of access control of multiple related, yet independent, software systems. With this property, a user logs in with a single ID and password to gain access to a connected system or systems without using different usernames or passwords, or in some configurations seamlessly sign on at each system. This is typically accomplished using the Lightweight Directory Access Protocol (LDAP) and stored LDAP databases on (directory) servers. A simple version of single sign-on can be achieved over IP networks using cookies but only if the sites share a common DNS parent domain.

[8] In cryptography, a side-channel attack is any attack based on information gained from the physical implementation of a cryptosystem, rather than brute force or theoretical weaknesses in the algorithms (compare cryptanalysis). For example, timing information, power consumption, electromagnetic leaks or even sound can provide an extra source of information, which can be exploited to break the system. Some side-channel attacks require technical knowledge of the internal operation of the system on which the cryptography is implemented, although others such as differential power analysis are effective as black-box attacks.

Exercises

[Ex. 1] Answer the following questions according to the text.

1. What is cloud computing security? What does it refer to?
2. What are the security concerns associated with cloud computing?
3. What should an efficient cloud security architecture do?
4. What does deterrent controls do? And how?
5. What does preventive controls do? And how?
6. What are detective controls intended to? What will a detective control do in the event of an attack?
7. What does corrective controls do? And how? When do they come into effect?
8. What do cloud providers do about identity management?
9. What are the security threats associated with cloud data services mentioned in the text?
10. What are the security requirements to be met in a cloud data service to throttle the threats?

[Ex. 2] Translate the following terms or phrases from English into Chinese and vice versa.

1. cloud computing	1. ______
2. insider attack	2. ______
3. outsourced data	3. ______
4. side channel attack	4. ______
5. public cloud	5. ______

6. 可控性	6. ______________
7. 尺度，维(数)，度(数)；元	7. ______________
8. 超级管理程序	8. ______________
9. 冲突，干涉	9. ______________
10.隔离，分离	10. ______________

[Ex. 3] Translate the following passage into Chinese.

Big Data is the current buzzword in the technology sector, but in fields such as security it is much more than this. Businesses are starting to bet strongly on the implementation of tools based on the collection and analyzing of large volumes of data to allow them to detect malicious activity. What started out at a fashionable term has turned into a fundamental part of how we operate.

So, what exactly are the advantages of big data? Well, have a think about the current situation in which the use of mobile devices is growing, the Internet of Things has arrived, the number of Internet users is reaching new highs, and quickly you realize that all of this is prompting an increase in the number of accesses, transactions, users, and vulnerabilities for technology systems. This results in a surge in raw data (on the World Wide Web, on databases, or on server logs), which is increasingly more complex and varied, and generated rapidly.

Given these circumstances, we are encouraged to adopt tools that are capable of capturing and processing all of this information, helping to visualize its flow and apply automatic learning techniques that are capable of discovering patterns and detecting anomalies.

[Ex. 4] Fill in the blanks with the words given below.

monitoring	exposure	relieves	reputational	mobile
requirements	organization	storage	independently	growing

Key Challenges for Big Data Security

Cyber Criminals. As it becomes bigger and more difficult to manage, big data consequently becomes more appealing to hackers and cyber criminals. Because big data is a dataset of unprecedented size with centralized access, any ___1___ is total exposure. These types of breaches make headlines, incite consumers, and may cause major ___2___, legal, and financial damage.

Resource Capacity. As an organization collects big data across channels at an exponential rate, their ___3___ can grow beyond terabytes. As a result, data encryption and migration can get bottle-necked or leaky. Additionally, the sheer volume of data makes implementation of security control unwieldy. The tools required for ___4___ and analyzing big data produce massive amounts of their own security-related data every day, which puts undue pressure on the organization's capacity to store and analyze it all.

Cloud and Remote Access. One answer to the capacity issues of big data is to put it in the cloud. This ___5___ some of the burden for storage and processing, but creates new challenges for protecting it from criminals. And as more businesses allow for flex-time and ___6___ offices, employees have access to sensitive company data via smart phones, tablet devices, and home laptops.

Protecting personal devices becomes a balancing act between security and productivity.

Supply Chain and Partner Security. Organizations rarely operate ___7___. They rely on supply chain partners and external vendors for many of their business functions. Information flows in and out of each ___8___ to keep these relationships functioning. Coordinating the safety of big data across partners is another layer of complexity to a business's information security challenges.

Privacy. Both private and public organizations face the ___9___ challenge of privacy concerns. Consumers are wary about personal information being collected and stored, and fearful about security breaches. Plus, there are legislative and regulatory ___10___ to keep in mind.

Text B

How Big Data Can Secure User Authentication

Password-based as well as two-factor and multi-factor authentication processes have not been able to provide protection to systems and data due to various reasons. Password-based authentication is too weak, and two-factor and multi-factor authentication processes have been rejected by users because of poor user experience.

Big-data-based authentication systems promise to offer both robust authentication and a good user experience. Unlike other authentication systems, big-data-based authentication authenticates a user based on multidimensional and regularly updatable information collected about the user. The main difference between big-data-based authentication and other processes is that the former uses multidimensional information to authenticate a user. Multiple such products are already available on the market, and they are becoming popular. However, other systems have not been consigned to oblivion just yet due to various reasons.

1. Current Trends in User Authentication

In the user authentication domain now, the traditional systems such as password-based systems are still being used, while novel methods such as big-data-based authentication are emerging. Traditional systems, for all their problems, are still being used because of lesser acceptance of stronger authentication systems and integration issues with newer models. Some of the main trends in this domain are described below.

- Many companies offer a combination of password-based and multi-factor authentication system, but the latter is optional for the users because many users find it inconvenient.
- Two-factor and multi-factor authentication, though better than a password-based system, have had limited adoption because of poor user experience.
- Many companies are using passive biometrics in which data about the user such as fingerprints, voice and face recognition[1] are collected and used to authenticate the user.
- Big data authentication is becoming popular because just like biometric authentication's

approach, it collects data about users and builds a profile of the user without the user knowing about it. The profile is regularly updated and used to authenticate the user.

2. How the User Authentication Process Works

For all the innovations in this industry, the core principle of authentication systems remains the same: Match user inputs with the available data in the system. The different authentication systems are described below.

- In the password-based system, the password provided by the user is usually matched with that stored in the database in an encrypted format earlier.
- In the multi-factor system, the system matches multiple passwords—some of which are stored in the database and the remaining dynamically generated—with the inputs provided during the access request.
- In the biometric system, the system collects data from a person's voice, fingerprints or iris and uses that data to authenticate the user.
- In the big-data-based system, the system creates a profile of the user based on the data it regularly collects. It authenticates access requests by matching access inputs with the data in the profile.

3. Challenges in the Current Process

The main challenges in the current process are described below.

- Organizations have been facing a lot of financial and technical challenges in moving from purely password-based systems to more secure authentication systems. For example, in a huge enterprise with a lot of legacy systems, migrating from one process to another could be a nightmare.
- Multi-factor systems tend to mar user experience and users tend to avoid layered authentication, if given an option. It is a challenge both getting users to follow the process and keeping the authentication system robust.

4. How Big Data Authentication Works

Big-data-based authentication systems create profiles of all valid users of a system based on data collected about the user. The user does not even know that the system has been collecting data. Whenever a request to access the system is sent, the authentication system matches the information collected when the access request was made with that in the profile. Any mismatch or deviation from the profile could set off a warning about unauthorized attempts.

Given the evolving nature of attacks, the big data authentication system performs pretty complex functions. According to Don Gay, the chief security strategist of a user behavior analytics company, "With bad actors increasing the sophistication of their attacks, enterprises are having a difficult time pinpointing the threats and vulnerabilities that pose the largest risk." The user data it collects can be varied, unstructured and complex, such as the following:

- Information-entering behavior: Does the user use a physical keyboard or a virtual keyboard provided on the website?
- What level of security permissions does the user have?

• How many attempts does the user normally take to enter the correct password?
• How many times on average does the user access the system in a day?
• How many times in the past has the user reset the password?

The system simultaneously collects data about the user and monitors his activities, too. The system has to adapt to the unique behavior of each user. As Ivan Tendler, the cofounder and CEO of Fortscale, a reputed user behavior analytics company says, "We look at this from the user's perspective. He has a name, a personality and habits. This user is sloppy or this user is risky or this user tends to have too much permission and so on. You have to look at the user history and profile his behavior. And only in those methods can you spot odd behavior and can pinpoint malicious users or compromised users whose credentials were stolen."

The authentication system collects large volumes of both structured and unstructured data[2] from a variety of sources and is able to analyze them, detect patterns of behavior and anomalies and detect attacks from a variety of sources such as network devices, security appliances, hosts, endpoints, applications and databases.

Organizations have been reaping benefits of this approach already. For example, the New Jersey Department of Labor and Workforce Development (NJDLWD) uses a big-data-authentication solution to identify fraudulent unemployment benefit claims. The data authentication system works in two steps: First, it establishes whether the identity presenting a claim is real, and second, whether the identification is owned by the person making the claim.

5. Future Trends

The following trends could possibly unfold:

• Password-based systems will be used in conjunction with other newer authentication systems.
• More investment will be made into making the user experience of two-factor and multi-factor systems better.
• Organizations will invest a lot into making biometric systems more acceptable and robust by addressing the limitations of voice-based authentication systems. It seems that iris-based authentication is going to find many takers.

6. Conclusion

Big data authentication is still evolving and it will be a while before more is known about the system and its acceptability in the industry. Theoretically, it sounds promising, though. For all its fragility, the password-based system will not be junked, but used in conjunction with other authentication systems such as the two-factor and multi-factor systems. Another factor that needs to be considered is the ability or affordability on the part of organizations to migrate from basic authentication systems to more robust and stable systems.

According to Gartner, many organizations have been finding it tough to incorporate advanced authentication systems into their systems. Many organizations will watch the developments on the big data authentication front with both interest and caution. This applies especially in industries that deal with a lot of confidential data such as banking and finance, defense and health care.

New Words

multidimensional	*adj.* 多面的，多维的
regularly	*adv.* 有规律地，有规则地；整齐地，匀称地
updatable	*adj.* 可更新的
novel	*adj.* 新奇的，新颖的
integration	*n.* 综合
trend	*n.* 倾向，趋势
	v. 伸向，倾向，通向
inconvenient	*adj.* 不便的，有困难的
passive	*adj.* 被动的
innovation	*n.* 改革，创新
iris	*n.* 虹膜
profile	*n.* 用户配置文件
purely	*adv.* 纯粹地，完全地
migrate	*v.* 迁移，移动，移往
nightmare	*n.* 梦魇，噩梦；可怕的事物
mar	*v.* 弄坏，毁坏，损害
	n. 损伤，毁损，障碍
mismatch	*v.* 使配错，使配合不当
	n. 错配，失谐
strategist	*n.* 战略家
sophistication	*n.* 世故，老练，精明
pinpoint	*v.* 查明
varied	*adj.* 各式各样的
unstructured	*adj.* 非结构化的，未组织的
cofounder	*n.* 共同创办人，共同创始人
personality	*n.* 个性，人格
sloppy	*adj.* 懒散的，草率的
spot	*v.* 认出，发现
odd	*adj.* 临时的，不固定的
endpoint	*n.* 端点，终点
reap	*v.* 收割，收获
unfold	*v.* 打开，显露，展开，阐明；呈现
acceptability	*n.* 可接受性
fragility	*n.* 脆弱，虚弱
affordability	*n.* 可购性，支付能力
stable	*adj.* 稳定的
development	*n.* 发展

caution	*n.*	小心，谨慎；警告
	v.	警告

Phrases

password-based system	基于密码的系统
big-data-based authentication	基于大数据的认证
user experience	用户体验
be consigned to oblivion	被遗忘，置于脑后
face recognition	人脸识别
available data	现有数据，现有资料
legacy system	旧系统，遗留系统，已有系统
layered authentication	分层认证
set off	引起
structured data	结构化数据
unstructured data	非结构化数据
unemployment benefit	失业救济
in conjunction with…	与……协力
voice-based authentication	基于语音的认证
iris-based authentication	基于虹膜的认证

Abbreviations

CEO (Chief Executive Officer)	执行总裁，首席执行官

Notes

[1] A face recognition system is a computer application capable of identifying or verifying a person from a digital image or a video frame from a video source. One of the ways to do this is by comparing selected facial features from the image and a face database.

It is typically used in security systems and can be compared to other biometrics such as fingerprint or eye iris recognition systems. Recently, it has also become popular as a commercial identification and marketing tool.

[2] Unstructured data (or unstructured information) refers to information that either does not have a pre-defined data model or is not organized in a predefined manner. Unstructured information is typically text-heavy, but may contain data such as dates, numbers, and facts as well. This results in irregularities and ambiguities that make it difficult to understand using traditional programs as compared to data stored in fielded form in databases or annotated (semantically tagged) in documents.

Exercises

[Ex. 5] Fill in the following blanks according to the text.

1. Password-based as well as two-factor and multi-factor authentication processes have not

been able to provide protection to ____________________. The reasons are that ________________ is too weak, and two-factor and multi-factor authentication processes have ___________________.

2. Big-data-based authentication systems promise to offer both ____________________ and ____________________. The main difference between big-data-based authentication and other processes is that the former uses ____________________ to authenticate a user.

3. Many companies offer a combination of ______________ and multi-factor authentication system, but the latter is optional for the users because ____________________________.

4. Big data authentication is becoming popular because just like biometric authentication's approach, it ______________ and ______________ without the user knowing about it. The profile is regularly ______________ and ______________.

5. In the password-based system, the password provided by ______________ is usually matched with that stored ______________ in an encrypted format earlier.

6. Organizations have been facing a lot of ______________ and ______________ challenges in moving from purely password-based systems to _______________________.

7. Big-data-based authentication systems create __________________ of a system based on _________________________.

8. Whenever a request to access the system is sent, the authentication system _____________________ when the access request was made with that in the profile. Any mismatch or _______________________ could set off a warning about ___________________.

9. The authentication system collects large volumes of _______________________ from a variety of sources and is able to ________________, detect patterns of behavior and anomalies and detect attacks from a variety of sources such as network devices, ________________, hosts, endpoints, applications and databases.

10. Organizations will invest a lot into making _________________ more acceptable and robust by addressing the limitations of ____________________________. It seems that iris-based authentication is going to find many takers.

Reading Material

What You Need to Know about Security Intelligence[1] with Big Data

1. Amplifying[2] Security Intelligence with Big Data

Leading security intelligence solutions today rely upon a set of structured and semi-structured[3] data sources, including logs, network traffic and others, to provide the Security Operations Center with an on-going real-time view of their organization's security posture[4]. The metrics employed to

---

1 intelligence *n.* 智力，聪明，智能

2 amplify *v.* 放大，增强

3 semi-structure *n.* 半结构化

4 posture *n.* 状态，情况

evaluate solutions include the scale and speed of data that can be processed in real-time, pruning[1] the large set of raw data to a limited set of significant security incidents requiring the attention of the organization.

While security intelligence solutions do enable security analysts to explore the data and identify emerging threats or pinpoint new risk exposures, the focus is on employing an existing portfolio of threat and risk identifiers to enable real-time analysis for detection. While this approach is effective for monitoring and maintaining the cyber defenses of an organization as well as improving the response time to handle incidents, a new set of challenges are surfacing which requires security intelligence to be amplified with big data analytics.

2. Proactively Mitigating[2] Risk and Identifying Threats

As the organizational perimeter blurs[3] due to rapid market adoption of cloud and mobile technologies as well as consumer engagement in social networks, an organization cannot solely focus on defense. Rather the organization has to be more proactive[4] in mitigating risk and identifying threats.

Attackers are also employing more sophisticated targeted attack techniques such as social engineering, and spear phishing[5]. The attack methodologies are also adapting to current defensive approaches—attempting to either hide malicious activities among large amounts of innocuous[6] activity or disguise[7] the intent by appearing to be innocuous activity. Even current tumultuous[8] economic and social conditions are further motivating[9] new types of malicious behaviors.

3. The Need for Big Data and Big Data Analytics

Evolving security intelligence to meet the needs of the new security challenges requires big data and big data analytics.

Firstly, an organization needs to keep its traditional security data for longer periods of time to perform analysis on the data. Historical analysis has the potential of unearthing[10] longer running attack methods and identifies relapses[11] in security over time.

Secondly, data sources not traditionally employed for security can help an organization better qualify what assets and entities need to be protected and/or observed[12]. For example, identifying users who most often work with sensitive data, and systems that are critical to core business processes. Data sources such as e-mail, social media content, corporate documents, and web content may help add

---

1 prune *v.* 修剪

2 mitigate *v.* 减轻

3 blur *v.* 把(界线、视线等)弄得模糊不清

4 proactive *adj.* 积极主动的

5 spear phishing 鱼叉式网络钓鱼

6 innocuous *adj.* 无害的，无毒的

7 disguise *v.* 假装，伪装，掩饰

8 tumultuous *adj.* 喧嚣的

9 motivate *v.* 激发

10 unearth *v.* 掘出

11 relapse *n.* 复发，回复原状

12 observe *v.* 观察，观测

additional context to traditional security data but are predominantly[1] unstructured data.

Next, a variety of analytics can be performed to reveal security insights from these larger data sets and will require more processing time. This analysis will need to be done asynchronously[2] to the real-time analysis that traditional security intelligence specializes in. However, once the analysis is complete, the insights[3] have to be fed back to the real-time component to make the overall solution more effective over time.

Finally, a renewed emphasis needs to be placed on investigative[4] analysis that can initially be categorized as ad hoc before it is codified[5]. Given the specificity of an organization and its business ecosystem[6] this will be crucial for the security intelligence solution to gain contextual[7] awareness necessary for thwarting targeted attacks.

4. Six Categories of Use Cases

Security intelligence with big data solution will empower[8] an organization to address the needs of a changing security landscape. The following are categories of use cases where it can prove at least beneficial if not essential.

(1) Establish a Baseline[9]

Organization gains an understanding of its ecosystem, what needs to be defended or observed as well as formulating a risk profile enabling it to detect abnormalities[10].

Common Use Case Questions:

- Who are the attractive targets within my enterprise?
- Which applications and what data do we need to defend due to their sensitivity[11]?
- What is the normal behavior profile for users, assets, and applications?

(2) Recognize Advanced Persistent Threats

Organization gains awareness of a motivated or incentivized attacker who attempts to hide or disguise the attack as innocuous interactions, potentially over a long period of time (months, years).

Common Use Case Questions:

- Which assets within my organization are already compromised or are vulnerable?
- Which external domains may be the source of attacks?
- Are there any low profile network traffic elements that might signal an ongoing or imminent[12] attack?

---

1 predominantly *adv.* 占主导地位地，显著地
2 asynchronous *adv.* 不同时地，异步地
3 insight *n.* 洞察力，见识
4 investigative *adj.* 研究的，好研究的
5 codify *v.* 整理，编纂
6 ecosystem *n.* 生态系统
7 contextual *adj.* 文脉上的，前后关系的
8 empower *v.* 授权于，使能够
9 baseline *n.* 基线
10 abnormality *n.* 异常性
11 sensitivity *n.* 敏感，灵敏(度)，灵敏性
12 imminent *adj.* 即将来临的，逼近的

(3) Qualify Insider Threats

Organization gains evidence or is warned of users within the organization's network who may be inclined to steal intellectual property[1], compromise enterprise systems or perform other actions that are detrimental[2] to the organization's operations.

Common Use Case Questions:

- What data is being leaked or lost and by whom?
- Who internally has the motivation[3] and skills to compromise the cyber operations of the company?
- Who is exhibiting abnormal usage behavior?

(4) Predict Hacktivism

Organization is alerted to attack from groups or entities that sympathize with causes that are contrary to the business interests of an enterprise.

Common Use Case Questions:

- Which controversial[4] issues may trigger a negative sentiment about the organization triggering an increased risk of attack?
- How to identify and monitor intentions of entities antagonistic[5] to the organization's business practices?
- How does publicity of the company in the media impact risk?

(5) Counter Cyber Attacks

Organization is informed of[6] an impending or on-going attack by criminal enterprises or government funded or government sponsored groups.

Common Use Case Questions:

- What is the origin of an attack?
- Which hacking tools may be used and who is gaining access to them?
- Are their symptoms[7] of an attack underway[8] or being planned manifesting themselves as support issues?

(6) Mitigate Fraud

Organization is appraised of new or existing fraud methods that may compromise its compliance with regulations or cause significant losses to its financial operations.

Common Use Case Questions:

- How can the organization identify a fraudulent[9] activity?

---

1 intellectual property　知识产权

2 detrimental　*adj.* 有害的

3 motivation　*n.* 动机

4 controversial　*adj.* 争论的，争议的

5 antagonistic　*adj.* 反对的，敌对的

6 be informed of　听说，接到……的通知

7 symptom　*n.* 征兆，症状

8 underway　*adj.* 起步的，进行中的

9 fraudulent　*adj.* 欺诈的，欺骗性的

- Which users have compromised identities that may lead to fraudulent activity?
- Can well known fraud attempts have patterns that can either be detected or even anticipated?

参考译文

云计算安全

云计算安全性，或者更简单地说，云安全是计算机安全、网络安全和更广泛的信息安全的不断发展的子领域。它是指为了保护数据、应用程序和相关的云计算基础设施而部署的一整套广泛的策略、技术和控制。

1. 与云相关的安全问题

云计算和存储解决方案为用户和企业提供了在第三方数据中心存储和处理其数据的各种功能。组织以各种不同的服务模式（SaaS、PaaS 和 IaaS）和部署模式（私人、公共、混合和社区）来使用云。有许多与云计算有关的安全问题。这些问题分为两大类：云提供商（通过云提供软件、平台或基础设施即服务的组织）所面临的安全问题及其客户（在云上托管应用程序或存储数据的公司或组织）所面临的安全问题。但是，他们共担责任。提供商必须确保其基础设施安全，并保护客户的数据和应用程序，而用户必须采取措施加强其应用程序并使用强大的密码和身份验证措施。

当一个组织选择在公共云上存储数据或主机应用程序时，它失去了对承载其信息的服务器进行物理访问的能力。因此，敏感数据可能面临内部攻击的风险。根据云安全联盟最近的一个报告，内部攻击是云计算的第六大威胁。因此，云服务提供商必须确保对具有物理访问数据中心服务器的员工进行彻底的背景检查。此外，必须经常监视数据中心以发现可疑活动。

为了节省资源、降低成本并保持效率，云服务提供商通常在同一台服务器上存储多个客户的数据。因此，有可能一个用户的私人数据可以被其他用户（甚至是竞争对手）查看。为了处理这种敏感的情况，云服务提供商应确保正确的数据隔离和逻辑存储隔离。

虚拟化在实施云基础设施方面的广泛使用为客户或公共云服务的租户带来了独特的安全隐患。虚拟化改变了操作系统和底层硬件之间的关系——无论是计算、存储还是网络。这就多了一层，即虚拟化。虚拟化也必须进行正确的配置、管理和保护。具体的担忧包括可能损害虚拟化软件或“管理程序”。虽然这些担忧在很大程度上是理论上的，但它们确实存在。例如，虚拟化软件管理软件的管理员工作站的漏洞可能导致整个数据中心不能正常运行或被攻击者按照自己的喜好来重新配置。

2. 云安全控制

云安全架构只有在实施了正确防御措施时才有效。高效的云安全架构应该认识到安全管理会出现的问题。安全管理通过安全控制来解决这些问题。这些控制措施是为了保护系统、减少攻击的影响。虽然云安全架构背后有许多类型的控件，但它们通常属于以下类别之一：

（1）威慑控制

这些控件旨在减少对云系统的攻击。像栅栏或财产上的警告标志一样，威慑控制通常会告知潜在的攻击者，如果继续进行，将会对他们造成不利后果，从而降低威胁级别。（有些人认为它们是预防控制的一个子集。）

（2）预防控制

预防控制加强了系统对事件的防范，其目的一般是减少漏洞，而不是实际消除漏洞。例如，对用户的强认证使得未经授权的用户不太可能访问云系统，而且云用户也能正确识别。

（3）侦探控制

侦探控制旨在检测并对发生的任何事件做出适当的反应。在发生攻击的情况下，侦探控制将指示预防或纠正控制来解决问题。通常使用系统和网络安全监控（包括入侵检测和预防安排）来检测对云系统和支持通信基础设施的攻击。

（4）纠正控制

纠正控制可以让事件的后果不那么严重，这会减少损害。它们在事件发生期间或之后生效。恢复系统备份以重建受损系统是纠正控制的范例。

3. 云安全维度

通常建议通过评估威胁、漏洞和影响来选择和实施与风险匹配的信息安全控制。

4. 安全和隐私

（1）身份管理

每个企业都将拥有自己的身份管理系统来控制对信息和计算资源的访问。云提供商或者使用联盟或 SSO 技术，或基于生物特征的身份识别系统将客户的身份管理系统集成到自己的基础设施中，或者提供自己的身份管理解决方案。例如，CloudID 为此问题提供了基于保护隐私的云端和跨企业生物识别解决方案。它将用户的机密信息与其生物特征联系起来，并以加密方式存储。利用可搜索的加密技术，在加密域中执行生物特征识别，以确保云提供商或潜在攻击者无法访问任何敏感数据，甚至无法访问各个查询的内容。

（2）物理安全

云服务提供者对 IT 硬件（服务器、路由器、电缆等）提供物理保护，防止未经授权的访问、干扰、盗窃、火灾和洪水等，并确保必备供给（如电力）足够，以尽量减少中断的可能性。这通常通过世界级（即专业指定、设计、构造、管理、监控和维护）数据中心提供云应用程序来实现。

（3）人员安全

与云服务相关的 IT 人员以及其他专业人员的各种信息安全问题通常通过雇佣活动的前期、中期和后期来处理，例如对潜在招聘人员进行安全检查、培养安全意识以及专业培训。

（4）隐私

提供商确保所有关键数据（如信用卡号）被屏蔽或加密，并且只有授权用户可以访问全部数据。此外，数字身份和凭证必须受到保护，就像提供商收集或生成的有关云中客户活动的数据一样，也必须加以保护。

5. 数据安全

云数据服务有许多安全威胁，不仅包括传统的安全威胁（如网络窃听、非法入侵和拒绝服务攻击），还包括特定的云计算威胁（如边信道攻击、虚拟化漏洞以及滥用云服务）。为了遏制威胁，云数据服务需要满足以下安全要求：

（1）数据机密性

数据机密性是指不向非法用户提供或披露数据内容的属性。外包数据存储在云中，且不受数据所有者直接控制。只有经过授权的用户才能访问敏感数据，而其他用户不得获取数据的任何信息。同时，数据所有者希望充分利用云数据服务（如数据搜索、数据计算和数据共享），而不会将数据内容泄露给其他对手。

（2）数据访问可控性

访问可控性意味着数据所有者可以选择性地限制对外包到云端的数据的访问。合法用户可以由所有者授权访问数据，而其他用户无权访问。此外，期望对外包数据执行精细化访问控制，即不同的用户应被授予对不同数据片段不同的访问权限。在不受信任的云环境中，访问授权只能由其所有者控制。

（3）数据完整性

数据完整性要求维护和保证数据的准确性和完整性。数据所有者总是期望其云中的数据可以正确和可靠地存储。这意味着不得非法篡改、不当修改、故意删除或恶意制作数据。如果任何不良操作损坏或删除数据，所有者应该能够检测到损坏或丢失。此外，当外包数据的一部分损坏或丢失时，用户仍然可以恢复数据。